Processing and Packaging of
Heat Preserved Foods

Processing and Packaging of Heat Preserved Foods

Edited by

J.A.G. REES
and
J. BETTISON,

CMB Packaging Technology,
Wantage,
Berks.

Blackie
Glasgow and London

Published in the USA by
avi, an imprint of
Van Nostrand Reinhold
New York

Blackie and Son Ltd
Bishopbriggs, Glasgow G64 2NZ
and
7 Leicester Place, London WC2H 7BP

Published in the USA by
AVI, an imprint of
Van Nostrand Reinhold
115 Fifth Avenue
New York, New York 10003

Distributed in Canada by
Nelson Canada
1120 Birchmount Road
Scarborough, Ontario M1K 5G4, Canada

16 15 14 13 12 11 10 9 8 7 6 5 4 3 2 1

British Library Cataloguing in Publication Data
Processing and packaging of heat preserved foods.
1. Preserved foods
I. Rees, J.A.G.   II. Bettison, J.
664.028

ISBN 0-216-92908-3

Library of Congress Cataloging-in-Publication Data
Processing and packaging of heat preserved foods / [edited by] J.A.G.
Rees and J. Bettison.
    p.   cm.
Includes bibliographical references.
ISBN 0-442-30282-7
1. Canning and preserving.   I. Rees, J.A.G.   II. Bettison, J.
TP371.3.P76   1990
664'.0282—dc20   89-29393
                                                         CIP

Typesetting by Thomson Press (India) Ltd, New Delhi
Printed in Great Britain by Thomson Litho Ltd, East Kilbride, Scotland

Preface

Why, when we are almost at the end of the second century in which foodstuffs have been packaged and heat processed, do we need another book on the subject? The food industry is in a state of change; consolidation, regrouping and merger have removed the elements of specialisation and without doubt have accelerated the transition of food processing from an art to science and technology. There are many driving forces for change. Today new pressures in the market place from the retailer and consumer, together with the strong driving forces for safety and protection of the environment, demand the involvement of the whole of the food chain.

The packaging of foods represents a partnership between the food processor and the package manufacturer. The subject is not taught in the same way as the other elements of science and technology that are involved in the food chain, yet today it represents big business and, in the case of heat-preserved foods, is an integral and indispensable part of the total process.

This book will introduce the major food packages, processes and good manufacturing practices to the reader. It aims to transfer knowledge and the experience of many years of practice (in the case of metal containers, the details of manufacture and surface coating are unique to this publication). It also looks to the future, bringing together, for the first time in one volume, those elements critical to the heat preservation of foods.

The book is aimed at the new graduate entrant to the food industry and those production personnel engaged in good manufacturing practice, quality control and engineering. Personnel employed by the material suppliers (for example, in the steel and plastics industries) and the materials converters (the packaging industries) will find that the book provides a useful insight into the technologies involved in heat preservation. The prime objective of the book, however, will be to direct the reader to the delivery of safe and nutritious food to the consumer.

J.A.G.R
J.B

Acknowledgements

I would like to express my sincere thanks to the authors who have contributed to this book, and to their respective organisations, without whose co-operation the manuscripts would probably never have materialised. I would like to thank Mr B.J. Rushbridge for his contribution on 'Food Contact' in the introductory chapter. I am also indebted to Mr M.G. Alderson for his encouragement, advice and constructive comments in the preparation of the manuscript. Last, but not least, I would like to express my appreciation to Mrs June Tranfield for her assistance and secretarial support, coping with the numerous amendments and manuscript changes an endeavour of this magnitude produces.

Contributors

J. Bettison CMB Packaging Technology, Denchworth Road, Wantage OX12 9BP, UK.

K.L. Brown Campden Food and Drink Research Association, Chipping Campden, Gloucestershire GL55 6LD, UK.

M.N. Hall Campden Food and Drink Research Association, Chipping Campden Gloucestershire GL55 6LD, UK.

K.A. Ito National Food Processors Association, 6363 Clark Avenue, Dublin, California 94568, USA.

R.J. Pither Campden Food and Drink Research Association, Chipping Campden, Gloucestershire GL55 6LD, UK.

P.J.G. Proffit Neste Chemicals International, Bazellan 1, B-1140, Brussels.

J.A.G. Rees CMB Packaging Technology, Denchworth Road, Wantage OX12 9BP, UK.

P.S. Richardson Campden Food and Drink Research Association, Chipping Campden, Gloucestershire GL55 6LD, UK.

B.J. Rushbridge CMB Packaging Technology, Denchworth Road, Wantage OX12 9BP, UK.

J.D. Selman Campden Food and Drink Research Association, Chipping Campden, Gloucestershire GL55 6LD, UK.

K.E. Stevenson National Food Processors Association, 6363 Clark Avenue, Dublin, California 94568, USA.

R.H. Thorpe Campden Food and Drink Research Association, Chipping Campden, Gloucestershire GL55 6LD, UK.

T.A. Turner CMB Packaging Technology, Denchworth Road, Wantage OX12 9BP, UK.

Contents

3 Heat processing equipment   50
P.S. RICHARDSON and J.D. SELMAN

4 Aseptic processing and packaging of heat preserved foods   72
K.E. STEVENSON and K.A. ITO

5 Packaging of heat preserved foods in metal containers   92
T.A. TURNER

6 Packaging of heat preserved foods in glass containers
J. BETTISON

138

9   The effect of heat preservation on product quality   221
M.N. HALL and R.J. PITHER

10   Recommendations for the good manufacturing practice of heat preserved foods   238
J.A.G. REES

Index   247

1 Introduction

J.A.G. REES

1.1 Current market size

Over £1500 million worth of canned food is bought in the United Kingdom every year [1]. Over 7.5 billion cans are consumed, 1.5 billion of which are imported into the United Kingdom (Table 1.1). In the United States, the value of heat processed shelf stable foods in 1987 was approximately $18 billion. This figure includes cans, glass jars and aseptic cartons and was supplied by a census of over 1100 establishments of US manufacturers [2].

These figures demonstrate the continued popularity of heat processed foods and in particular the can. Contrary to what some may believe, the sales of canned food are forecast to show continued growth over the next decade with fruit, vegetable and fish products likely to show the greatest increases.

In addition to the traditional established forms of heat processed containers, the can and glass jar, new ones are emerging. Plastics containers have been commercially introduced, both in the United States and Europe for ready meals (Figures 1.1, 1.2). One significant success in the United States has been a 4 oz container for aseptically packed apple sauce; 500 million units were packed in 1988 (Table 1.2) [3]. In the United Kingdom, sales of a 125 g container for pet food have risen from zero to tens of millions in less than 2 years.

1.2 Consumer trends

Entering the 1990s, consumers are demanding foods which they perceive as being

- Better nutritionally
- More 'natural'
- Offering more convenience, better suited to fast moving life styles
- Safe, with high integrity

Better nutrition was associated initially with lower calories. The food manufacturers are now addressing the consumers' need for more specific health requirements. Higher fibre content, lower fat, low sugar, low salt, no additives or preservatives etc. are all featured in heavy promotional advertising.

Table 1.1   Current value of UK canned foods, 1987.

Market sector	Value (£ million)	Percentage of market
Meat products	280	18.5
Vegetables	238	15.5
Baked beans	195	12.9
Fruit	192	12.7
Fish	186	12.3
Soup	168	11.0
Milk/desserts	122	8.0
Pasta	78	5.1
Cooking sauces	62	4.0

Source: [1].

Figure 1.1   Shelf stable heat processed products (Plastic Containers, Europe).

'Natural' is usually associated with 'fresh' which in turn is usually associated with short shelf-life, perishable foodstuffs. 'Organic' foods i.e. foods grown without chemical fertilisers are starting to appear in our retail outlets.

The need for more convenient packaged foods has seen the growth of products targeted specifically at the domestic microwave oven. The penetration of these ovens in homes in both the United States and Europe has been rapid and continues to grow. In 1989, penetration of microwave ovens in US households reached 80%; sales of prepared and packaged foods targeted for the microwave oven exceeded $7 billion and are predicted to rise to $60 billion by the end of the century. One of the significant product growth areas has been in the development of ready meals in high barrier plastic trays specifically aimed at microwave reheating in the home.

Figure 1.2 Shelf stable heat processed products (Plastic Containers, USA).

Table 1.2 Multi-layer rigid barrier packaging market, 1988.

Application	Package	End users (partial)	MM units	Material use (MM lb)
Apple sauce	4 oz PP/EVOH or styrene/PVDC	Motts Musselman's Hunt-Wesson	500	6.5
Fruit juice	6 oz styrene/ PVDC	Ocean Spray Wyler's Slush	100	1.0
Puddings, sauces	4 oz PP/EVOH or PP/PVDC	Hunt-Wesson Kraft (Cheez-Whiz)	300	3.5
Entrées (retorted)	8 oz PP/EVOH bowls	Dial 'Lunch Bucket' Hormel AHF (Chef Boyardee)	100	4.0
	8 oz PP/EVOH cans	Hormel Ross Labs	25	1.0
	9–13 oz PP/EVOH or PP/PVDC trays	Hormel (Top Shelf) Del Monte Campbells Magic Pantry	100	5.5
Total			1125	21.5

Source: [3].

Fast food restaurants have achieved rapid growth again reflecting the attempt to satisfy our fast moving life styles. The scientist and technologist should always remember, however, that while we should respond to the needs and challenges of the market, safe processing and packaging remain the prime requirements.

The potential for food poisoning is of major concern. In Europe recently, problems of *Salmonella* contamination in eggs and *Listeria* in chilled cooked patés and sliced meats have focused attention on the risks to health. Canned foods have been associated with food poisoning but fortunately the incidence has been very low. The consumer will continue to demand that the industry delivers absolutely safe products.

1.3   Function of the package

The traditionally accepted functions of the package are that it should contain, protect, inform and attract. Today cost factors and legislative and environmental constraints are of paramount importance and to the traditional functions must be added those of environmental acceptability while meeting all of these requirements at minimum cost.

For most food products the over-riding objective is to ensure that the package will provide the optimum protective properties to preserve the product it contains in good condition for the anticipated shelf-life.

Physical and mechanical protection for the product must be afforded by the package to prevent damage, infestation, contamination, moisture pick-up, etc. For all methods of food preservation the package is an integral part of the process. In thermal processing, microbiological protection must also be afforded through the provision of a container of high integrity with a hermetic closure.

Container/product compatibility is of extreme importance. The specification for the package, while retaining physical and microbiological integrity, must not bring about any deterioration in the organoleptic characteristics of the contained food or obviously endanger the public health. These requirements are now mandatory in the United Kingdom by 'The Materials and Articles in Contact with Food Regulations 1987'.

Quality management procedures are therefore vital in the initial selection and subsequent use of container specifications for food packaging. Different packaging materials have different significant components, whilst the food and the environment in which it is sold and distributed may be similar. With metal containers, the selection of the correct container specification is critical to minimise corrosion with subsequent trace metal pick-up by the product. Similarly container and closure construction is important to prevent leakage, both physical and microbiological.

Flexible containers have detailed specifications for laminate performance and seal integrity, together with strict definitions for the amount of extractable material permissible under specific test conditions.

Rigid plastic containers again have to conform within fairly tight limits regarding plastic monomer content. For specific uses, e.g. where they are used for the packaging of dairy products, the empty containers have to be manufactured under strict conditions of hygiene so that they are supplied to the industry with good microbiological integrity.

Packaging for the food industries has therefore to satisfy many requirements which are subject to quality control during manufacture. The assurance of this quality is essential to the growth of both the packaging and food industries.

1.4 Selection of the package

The food packaging and processing industries are dynamic, constantly evolving and developing in response to market needs, increasing raw material costs and environmental pressures. The choice of container, food processing and preservation process is increasing. How then, when required to make selections, may the reader make the optimum choice? Experience leads one to the fact that no decision has ever been made objectively, since there are varying wants, needs, wishes and desires which influence decision. Package selection is after all a compromise between, in basic terms, performance and cost.

In the Western world we all have enough food to exist. We enjoy regular supplies of high quality foods and the spending of significant levels of discretionary income is common in our lifestyles. New product development will be influenced by demographic trends, 'fashion', etc. It should always be remembered, however, that the consumer's perception of quality and value for money is the only criterion for successful establishment of a new product.

Consumer perception is, as previously discussed, paramount. Assuming a new product has been developed, how do we select the optimum route through to production?

1.4.1 *Choice of container*

The choice of container will largely be dictated by:

- Ease of handling
- Speed of filling
- Ease of closing
- Ease of processing
- Shape/design
- Printing/labelling
- Required shelf-life

- Usefulness
- Consumer need
- Regulatory/environmental requirements

The various performance attributes of cans, glass jars and plastics containers are discussed in this book. If containers are required for microwave reheating in the home, shape and design are significant with regard to effective and even temperature distribution during re-heating. For optimum microwave reheating, containers should be wide-necked in the case of glass jars, or have a large, exposed surface area in the case of plastic trays. Containers should be shallow and corner design is important to facilitate air removal as the product heats up thus preventing the phenomenon known as the 'volcano effect', which creates spurting.

Shelf-life is determined by package composition, product/container compatibility and temperature of storage. Each container system has its own particular merits and de-merits which are discussed in the various chapters.

For metal containers, internal corrosion and storage temperature are the most important considerations; glass jars are inert and are impermeable but where light has an influence on product shelf-life then obviously glass is at a disadvantage. For plastic containers, gas, particularly oxygen, transmission is the most significant feature. Many products are found, via test pack performance, to be less oxygen sensitive than imagined. This appears to be particularly true of some ready meals and some dairy products. It may be possible to pack many of these products in low barrier or even non-barrier containers, particularly where shelf-lives of 3 months or less are all that is required. For citrus juices, the temperature of storage (0–5 °C) appears to be more significant than oxygen pick-up in achieving a shelf-life of up to 6 months. Shelf-life prediction is extremely difficult to make. There is no substitute for pack test experience.

Technical choice will also be influenced by plant investment and the filling, closing, conveying and heat processing facilities available. Most food manufacturers will obviously seek to obtain maximum efficiency of use and cost effectiveness from their existing investment.

Introduction of new products and/or new containers therefore should be considered in relation to the food manufacturer's processes. The product and container will obviously be of more benefit where little or no alteration to the existing manufacturing processes needs to be made. The introduction of the plastic bowl-shaped container for snack meals in the United States and more recently in the United Kingdom was facilitated by the fact that it could be filled and closed on existing equipment which needed only minor adjustment and hence little further capital expenditure on the production lines.

The technical choice will be made on the basis of closure integrity, ease of processing and required shelf-life. The merits and requirements for container closing, whether by double seaming, capping or heat sealing are discussed elsewhere.

Table 1.3 Process times for flat and cylindrical containers.

	Process time (min) to Fo10 lethality	Process medium
450 g barrier plastic tray, 'TOR' closed (hydraulically solid)	47	Water/air
450 g barrier plastic tray, gas flushed	53	Water/air
450 g cylindrical can	83	Steam

Source: [4].

For most heat preservation processes, the package is an integral part of the process, and is required to withstand the demands of the heating and cooling cycles. The rate of heat penetration into the product may be significantly affected by the container design. While the material of manufacture, metal, glass or plastic, will have an effect on rate of heat transmission, this effect is insignificant when the shape and profile of the container are taken into account. In general, the narrower the profile the better, i.e. process times for flat containers are shorter than for cylinders (Table 1.3).

1.4.2 Choice of process

The choice of heat process will be influenced by those product characteristics which the product development team most wish to preserve. We will assume that the achievement of commercial sterility is the first priority. Then texture, flavour, colour and nutrient retention will be considered. The technological inputs will be variable and the more complex the food product, the greater the technical hurdles. The longer the period between preparation and consumption, i.e. the shelf-life, the broader will be the scientific challenge.

Consumers are unlikely to be impressed by the technology used, in fact too much technology causes concern and suspicion, but they will be influenced by the package which delivers the product to their kitchen. A mistake often made by product development teams is that of attempting to reflect minor nuances or personal preferences in the formulation when what the consumer is looking for are major leaps in the perception of quality.

The decision to use conventional heat processing or aseptic processing will obviously depend on the plant installation, investment, etc. Aseptic processing has obvious advantages for the heat treatment of those foodstuffs that are more heat sensitive. It is important to remember that comparisons of identical product formulations processed by one method should not be made against another method. The product perceptions will be different.

Little is to be gained from seeking 'quality' improvements in products well established in the market place, where the perception is well identified by the consumer, by changing the method of processing and in many cases the packaging as well.

1.4.3 *Choice of market*

The technologist working in the food industry will always work closely with the marketeer such that identified trends and needs may be progressed efficiently. The marketeer will perceive the need for a new product or container where shape, design, decoration or some particular characteristic such as ease of use or opening, table-readiness, etc. will be featured. The technologist's brief will then be to deliver that product and package.

Microwavable containers, flexible, squeezable bottles for sauces and preserves, easy open 'ring-pull' ends for foods and beverages, shaped cans, and lightweight toughened glass jars are all examples where technological advance has been compatible with market requirements.

The demographic trends perceived for the 1990s are:

- Declining population of school age children
- Increasing population of mid-20 year olds
- Even greater population of people in their early 40s and 50s
- People over retirement age increasing
- Higher proportion of working women
- Increasing population of 'singles'

These trends together with the consumer issues already discussed, will dictate many of the directions in which food product and package development will progress.

Consumer affluence will also influence food choice. With more affluence, consumers become more adventurous in the foods they eat. As the demand increases, the food manufacturers have to move to larger scale production and distribution, creating opportunities for the development of food packaging systems.

Consumer choice has never been greater; in the United Kingdom, where the top five retailers account for 60% of all grocery sales, competition for shelf space between different food manufacturers is fierce. The marketeer will therefore seek to differentiate his product as much as possible from his competition. Raw, chilled, frozen or shelf-stable, canned, bottled, cartoned, each technology has a distinct consumer perception and a different package.

1.5 History and development

Trends in food production and consumption mirror the social and economic development of the population. From the end of the 14th century up to the beginning of the 18th century, the UK population had risen by only 3.5 million, from 2.5 to around 6 million. This slow growth reflects periods of war, recurring pestilence and poor harvests.

In the latter part of the 18th century, agrarian revolution led to the disappearance of open fields, the rotation of crops was further extended replacing the inefficient 'three field system', scientific methods were applied to agriculture and large farms superseded small-holdings. The population movement off the land and into towns which started in the late 18th century was made possible by such agrarian reforms, sustained by the growth of farm output in the United Kingdom.

In the 18th century, life expectancy was around 40 years and 60% of income was spent on food. As the 19th century progressed, industrial and agricultural output expanded with the population growing to 24 million. The food manufacturing industry responded to the growth in demand with advances in food technology such as pasteurisation and initial moves towards bottling, canning and freezing.

By 1900, with the population growing to over 38 million, increased industrialisation and the growth of cities had necessitated better transport and storage of food, with increased supplies of fresh fruit and vegetables available and growing supplies of imported food from overseas. Life expectancy was now over 50 years, with less than half of income spent on food.

Food preservation techniques were developed commercially, especially canning and quick freezing. The health of the population continued to improve and living standards rose. The amount of discretionary income grew, more money was available for spending on non-essentials. By 1985 approximately 20% of income was spent on food [5].

1.5.1 *Development of canned foods*

Until the late 19th century, the process of food preservation was far in advance of the knowledge of its scientific base. In 1860, Louis Pasteur explained the principle on which heat preservation is based after carrying out experiments related to sterilisation by heat, inhibiting the growth of micro-organisms and preventing re-contamination of the food contents through the provision of a hermetically sealed container.

This was some 50 years after the publications of Nicholas Appert describing methods for the preservation of foods. While Appert did not appreciate the need for microbiological inactivation, his ideas for cleanliness and 'quality control' were revolutionary in 1810. He instructed 'In general, all vegetables...must be picked as freshly as possible and treated with the greatest speed, from the garden to the cooker in a single bound.' This dictum well applies to today's industry.

Appert's book was reprinted in 1811 when an English translation appeared. The editors wrote, 'From the public papers we learn that a patent has been taken out for preserving provisions...described in this book.' This patent, taken out by an English merchant Peter Durand, followed Appert's book word for word in many passages! Durand's patent, however, first suggested the use of metal 'canisters', from the Greek 'Kanastron', a reed basket for carrying foods. The following year Donkin, Hall and Gamble purchased

Durand's patent for £1000 and the canning industry was born.

The Navy were first to appreciate preserved food as it provided a welcome alternative to their usual fare of salt meat and hard biscuits. An early reference is made in 1814 when Admiral Cochrane, commanding the West Indies Station, requested some 'patent preserved meats' to test on his sick sailors. Known as soup and bouilli, this became 'bully-beef' to the men unfamiliar with the pronunciation.

It was, however, the Arctic explorations which commanded much public interest and attention, and which supported the reputation of canned foods for maintaining a healthy diet. Following convincing reports from the Arctic voyages, preserved foods were carried by ships of the Royal Navy after 1831. It took time, however, for the acceptance of canned foods to reach the general public.

Cans of tomatoes, peas and sardines were available to the general public as early as 1830. They were, however, very expensive. Prices for soup, corned beef and salmon were the equivalent of 9.5–13 p, 11.5 p and 14.5 p, respectively when house rental was around 16.6 p per week. Sales were slow, not helped by the size of the cans (from 4 to 45 lb capacity) and convenience was not a priority when it needed a hammer and chisel to open the can. High prices stemmed from the laborious canning method; a good tin smith turned out ten cans a day by hand.

Developments in the canning industry were rapid in the second half of the 19th century. In 1847, the invention of a machine for stamping out can bodies was followed by machinery for the cleaning of food and cans and conveying them along processing lines.

In the United States, development had also been taking place. In 1819 an English immigrant, William Underwood, opened a factory in Boston preserving fruit, pickles and sauces. Thomas Kensett set up a similar factory in New York. Both Underwood and Kensett initially used glass jars and cork stoppers as originally reported by Appert.

The American Civil War caused an additional increase in demand for canned foods following earlier demands by shipping companies and as vital stores carried by wagon trains in the expansion of the West. In 1872, the Chicago meat canning industry was established and over the next 10 years, the canning of fish and vegetables grew significantly.

In 1874, A.K. Scriver of Baltimore invented the pressure retort. This enabled heating and cooling times to be considerably reduced. Donkin possibly used special vessels capable of being subjected to high pressure during his processing but the Americans first applied the process on a large scale.

Bacteriology was directly applied to food canning in 1895 when Brescott and Underwood carried out investigations into the spoilage of US canned corn, based on Pasteur's work.

Max Ams, a German settler in the United States in 1860, realised that the success of his canning business would depend upon the introduction of automatic methods. With a fellow countryman, Julius Brenzinger, a machine was invented for making double seams in can bodies. In 1897, Max Ams' son Charles patented a sealing gasket; this made the double seam air-tight and thus created an opening for high speed can manufacture. This 'Ams can' became very successful and in 1904 the Sanitary Can Company was formed through the combination of Ams, Cobb and Bogle.

The sanitary can became known as the open top can as the can and loose end were supplied to the food processor separately. The can and end were sealed after filling on the processor's premises. In 1908, the Sanitary Can Company was absorbed by the American Can Company.

Development in Britain also proceeded at a pace at this time. Great attention was paid to finding strains of fruit and vegetables which were best for canning, together with more efficient farming methods. In the 1920s and early 1930s, British packers made most of their own cans, unlike their American counterparts who used the open top can. A new branch of the industry was set up to provide canners with a supply of reliable food cans. In 1927, Williamson's of Worcester installed a Max Ams can line. In the first year, 3 million cans were produced. Production trebled the following year and 27 million were manufactured by year three. In 1930, Williamson's joined the Metal Box Company who acquired exclusive UK rights for 15 years to use the high speed machinery, methods, procedures and patents of the US Continental Can Company. The UK canning industry steadily progressed, became more efficient and canned foods became fully accepted as part of the national diet (Figure 1.3) [6, 7].

Figure 1.3 Range of early cans from 1824 (Parry's expedition), 1907 (Scott's expedition) and the 1940s (luncheon meat).

1.6 Materials and articles in contact with foodstuffs

1.6.1 *Introduction*

When considering the packaging of food, container/food product compatibility is of major importance. The close proximity, i.e. 'contact', of the foodstuff could lead to the transfer of constituents from the package material to the foodstuff. In some cases this could lead to a safety issue for the consumer. Protection of the consumer is the prime reason for laying down regulations and codes of practice appertaining to 'Food Contact'. 'Food contact', materials and articles in contact with foodstuffs, is therefore of increasing importance to the manufacturer in the food supply chain. It has moved to the forefront of legislation within the scope of international food law. The rapid introduction of new materials and packages to meet market demands only increases the need for awareness and action for the potential safety issue.

1.6.2 *Approaches to food contact*

Most countries have developed some form of control for food contact materials. One approach is through the use of Codes of Practice which relevant industries will help to create and will follow. This has been used in the United Kingdom with the 'Plastics for Food Contact Applications' Code issued by The British Plastics Federation. Another approach is through Statute Law, such as the United States' Food and Drug Administration, and their Code of Federal Regulations.

Whatever the local regulatory approach, the food contact issue is usually dealt with in one, or all, of the following ways.

- *Approved lists* Toxicological risks from the materials and the constituents of the materials would be assessed by an appropriate government/industry/academic body. The study would cover a range of toxicological aspects such as the state by which the constituents are bound up in the material, the risk of migration from the material and toxicological data derived by feeding trials, potential daily intakes, etc. Arising from the study of these toxicological aspects, the constituents or the material would be declared to be approved or not.
- *Migration tests* The extent by which a constituent of a material may be leached out into the foodstuff is very important. However, it is not necessarily a toxicity issue since the constituent may be safe but the level of extraction may be an indicator that the material is not inert enough for the task in hand.

Specific migration tests refer to named and specific constituents limited by defined maximum levels usually resulting from toxicological data. Overall or global migration tests measure the total amount of all the components extracted under the conditions of test. As such, this test is not in itself a measure of toxicity since it is an uncharacterised total extract of any number of components.

The measurement of trace amounts of chemical constituents in a complex biochemical mix such as a food stuff is very difficult. The mix of migrants, of unknown composition and quantity, masked by the composition of the food itself, presents an impossible task when measuring overall migration. A compromise is therefore necessary. Carefully selected simulants are used in place of the food, recognising the key constituents of food likely to cause potential extraction from the packaging material to occur e.g. acidity, fat and alcohol content. In each case, the regulations and codes will define the test conditions usually based on the conditions of use for that particular package and material combination.

1.6.3 Legislation

Legislation enacted in the United States under the auspices of the Food and Drug Administration has had the greatest impact. Many countries in Europe and the Far East have used the US regulations as a means of quickly establishing a basis for screening food contact materials. The FDA therefore has had a considerable influence as the benchmark, in some cases stimulating new legislation around the world.

United States   The FDA Code of Federal Regulations Title 21 are the food contact regulations. They provide approved or positive lists for a wide range of materials. The regulations are basically of two types. The first relates to components deliberately added to the material to make it suitable for certain performance criteria in package manufacture, e.g. plasticisers, heat-stabilisers, anti-oxidants, etc. The second relates to basic plastics polymers and covers aspects such as residual monomer content and molecular weight, overall migration limits, etc. The food category is defined, the condition of use categorised and the solvent to be used in the migration test is listed. The FDA Code of Federal Regulations Title 21, parts 170–199 provide all the details.

United Kingdom   As with all members of the EEC, UK regulations associated with food contact are in a state of change. Currently, the main cover is with Statutory Instrument 1523, October 1987 'Materials and Articles in Contact with Food'. This regulation carries a clear message and in effect is all embracing for every food contact situation.

Member States and EEC   The harmonisation of the Food Laws in the EEC is targeted for completion by 1992/3. The Legislation of a few of the Member States has become of wider value over the years and in the interim period before the EEC directives begin to take precedence, has been used within the

European arena as a source of reference. This is particularly so for countries such as Germany (Empfehlungen des Bundesgesundheitsamtes), Holland (Verpakkingen-en gebruiksartikelenbesluit), France (Matériaux au contact des aliments et denrées destinés à l'alimentation humaine, No. 1227).

The EEC harmonisation programme covers a broad spectrum of food contact materials and articles. An overall Framework Directive is in place (Amended Framework Directive 89/109/EEC) which lays down the basis for food contact legislation. As the legislative process speeds up most attention has been directed at plastics (adopted November 1989), regenerated cellulose film (83/299/EEC and 86/388/EEC) and ceramics (84/500/EEC). Metal packaging has not yet been directly involved. However 'metals and alloys' are on the EEC list for future action, and preparation is already under way for 'can coatings'. As with plastics, any directives for metal containers will need appropriate, suitable and practicable test methods.

References

1. *Canned Foods in the 1990s*, Canned Foods Information Centre, February 1988.
2. National Food Processors Association, pers. commun., January 1990
3. *Packaging Strategies*, USA, July 1989.
4. Alderson, M.G., *Trends in Food Packing System*, World Packaging Congress 1989, Singapore.
5. Annual Abstract of Statistics, H.M. Government Publication, 1988
6. Thorne, S., *History of Food Preservation*, Parthenon, 1986.
7. *History of Canned Food*, MB Publications, Reading.

Further reading

Sacharow and Brody, *Packaging*, Harcourt, Brace, Jovanovich, New York, 1987.
EEC Food Laws Symposium 1989, Publication No. 43, Food RA (UK) February, 1990.

2 Principles of heat preservation

K.L. BROWN

2.1 Introduction

It is important to understand what is being achieved by the application of a heat process. The primary objective may be to destroy micro-organisms capable of growing in the product at the intended distribution temperature or of endangering the health of the consumer. However, for a number of products, organoleptic properties may be more important in determining the severity of the heat process. Processes used in fish canning, for example, are designed to soften the bones of the fish resulting in an overkill with respect to the micro-organisms. Meat processes are also often designed to cook and tenderise the product resulting in a process that is more than adequate in terms of microbial stability. In the pasteurisation of pickles, the destruction of enzymes may be more important in determining the heat process than the destruction of micro-organisms in order to ensure a shelf-stable product.

For a growing range of products, the heat treatment is only part of a preservation process, and is used in combination with other factors, for example, lowered water activity, lowered pH, high salt content or lowered storage temperature. For products in this category, the heat process may only achieve a reduction in numbers of micro-organisms such that spoilage does not occur within the shelf-life of the product. The control of food poisoning organisms is always an important consideration and the destruction of particular pathogens which may or may not grow in the product may be the aim of the process. For example, *Salmonella* will not proliferate in dried products but it is important to ensure their destruction (by heat) to minimise the risk of infection of the consumer.

2.1.1 *Sterility*

The heat treatment of food is often mistakenly termed sterilisation. It is important to recognise that a product that has been subjected to heat 'sterilisation' may not be sterile. If micro-organisms are destroyed by heat at an exponential rate then it follows that there will be no absolute endpoint. The heat treatment is simply reducing the probability of survival. In practical terms, however, it is possible to reduce the probability of survival to such an extent that product may be regarded as 'sterile'.

2.1.2 Commercial sterility

'Commercially sterile' food may be defined as a product which has been processed so that under normal storage conditions it will neither spoil nor endanger the health of the consumer [1]. For example, an acid product such as fruit may have received a pasteurisation process sufficient to deal with yeasts, moulds and non-sporing bacteria but insufficient to destroy bacterial spores. With the exception of some aciduric species, the presence of viable bacterial spores in such high acid products is considered to be insignificant since the acidity prevents their development. Processors should exercise caution in using 'commercially sterile' acid foods (pH < 4.5) such as fruit as ingredients in new products (e.g. meat-based) where the high pH may allow the proliferation of spores of food poisoning organisms that have survived the pasteurisation given to the fruit. It is also important to control spoilage in acid products (pH < 4.5) since the spoilage organisms themselves may raise the pH, thus allowing pathogens such as *Clostridium botulinum* to develop. The pH of the food is an important factor in determining process severity and the pH values of various products including the work of Adam and Dickinson [2] are presented in Table 2.1.

2.2 Methods of determining heat resistance of micro-organisms

For a detailed explanation of the different techniques which have been used to determine the heat resistance of micro-organisms, the reader is referred to Pflug and Holcomb [3], Brown and Ayres [4], Hersom and Hulland [1] and Stumbo [5]. Examples of methods used are listed in Table 2.2.

The simplest method is probably the capillary tube technique where micro-organisms are sealed into capillary tubes which are then heated in a stirred oil bath. Similar techniques include the thermal death time tube, thermal death time can and plastic rod methods. These are all examples of indirect heating systems where there is a physical barrier (e.g. the wall of the capillary tube or can) between the heating medium and the suspension of micro-organisms. The temperature of the sample rises exponentially which means that a significant proportion of the heating up or come-up time is at the higher temperatures where lethality is greatest. This is the major limiting factor of indirect heating methods.

Indirect heating methods may also include continuous flow of product or suspension of micro-organisms through a heat exchanger (e.g. tubular or plate system). The heating profiles in these systems can be very complex, and difficult to model satisfactorily.

In mixing methods, a small volume of liquid containing micro-organisms is

Table 2.1 pH values of various products.

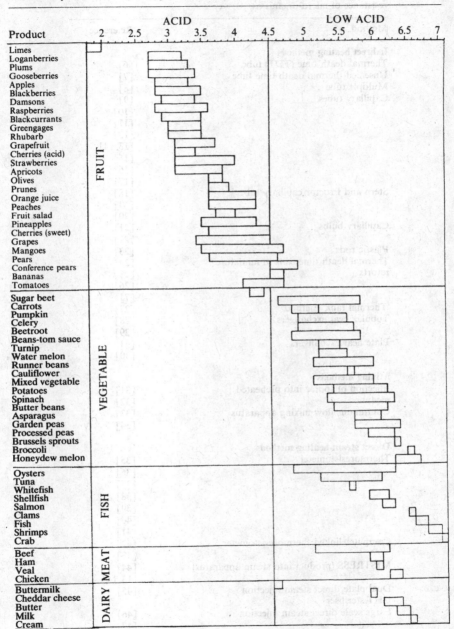

Table 2.2 Examples of methods used for determination of wet heat resistance of micro-organisms.

Method	Reference
Indirect heating methods	
Thermal death time (TDT) tube	[6]
Unsealed thermal death time tube	[7]
Multiple tube	[8]
Capillary tubes	[9]
	[10]
	[11]
	[12]
	[13, 14]
	[15]
	[16]
	[17]
Stern and Proctor capillary tube	[18]
	[19]
	[20]
Capillary bulbs	[21]
	[22]
Plastic rods	[23]
Thermal death time cans in miniature	[24]
retorts	[25]
	[26]
	[7]
'Thermal tank' method	[27]
Tubular heat exchangers	[28]
	[29]
Plate heat exchangers	[9]
	[30]
Mixing methods	
Injection of spores into preheated	[31]
medium	[32]
Continuous flow mixing apparatus	[33]
	[34]
Direct steam heating methods	
Thermoresistometer	[35]
	[36]
	[37]
	[38]
	[39]
	[40]
	[41]
Computer linked thermoresistometer	[42]
	[43]
MISTRESS (product into steam apparatus)	[44]
	[13, 14]
Dual plate/direct steam injection	[45]
UHT steriliser	
Large scale direct steam injection	[46]
	[47]
Alfa Laval Vacu-Therm instant	[48]
steriliser	

Table 2.2 (*contd.*)

Method	Reference
Particle methods	
Paper rolls in glass tubes	[49]
Perspex beads	[50]
Alginate beads	[51]
Alginate/food particles	[52]
Inoculated thread in cubes of product	[53]
Capillary bulbs in cubes of product	[54]
Electrical heating	
Differential scanning calorimeter	[55]

mixed with a much larger volume of substrate which has been preheated to the required temperature. Almost instantaneous heating can be obtained using these methods providing that the mixing takes place rapidly. By mixing a suspension of micro-organisms with steam, extremely rapid heating can be achieved due to the latent heat energy released when steam condenses. This forms the basis of the steam injection, thermoresistometer and MISTRESS methods.

Process times for particles are usually considerably longer than those for liquids and methods have been devised to immobilise micro-organisms inside food particles in order to ensure adequate processing. The alginate bead and alginate particle methods, for example, have been used successfully to determine process times for particle heating systems in which the particles are usually heated by the carrying fluid. In the Ohmic particle heating system (APV, Crawley), however, the particles heat up faster than the carrying fluid and again the alginate particle method can be used in this system to monitor bacterial destruction.

One of the more unusual methods involves heating suspensions of micro-organisms in small metal containers in a differential scanning calorimeter. In this apparatus, the cups are heated electrically, and by varying the power input, the rate of heating can be altered. Electrical heating usually generates a linear rise in temperature in the sample, although by varying the power input, a variety of heating profiles may be generated. There has recently been interest in the use of Peltier devices for heating micro-organisms. A Peltier device is, in effect, a thermocouple in reverse. By applying a dc voltage across a thermocouple, the junctions will heat or cool depending on the polarity.

2.3 Factors affecting the measurement of heat resistance

Apart from the obvious influences of suspending medium, pH, water activity, salt concentration, etc. the most important factor for consideration in any heat resistance experiment is the rate and mechanism of heat transfer and its

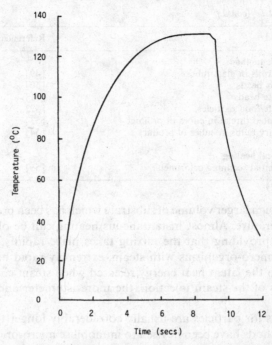

Figure 2.1   Temperatures on central axis of a capillary tube heated at 130°C in an oil bath.

measurement. The multiplicity of methods devised for determination of heat resistance reflect the attempts of different investigators to control and measure heat transfer. It may be pertinent here to list some of the problems that may have an influence on the results obtained.

As experimental temperatures are increased and exposure times become shorter, the limitations of different methods become apparent. Capillary tubes that have been used successfully for low temperature (below 130°C)/long time exposures (> 1 min) suffer from the drawback of rate of heat penetration at high temperatures. Perkin *et al.* [11] used thin wire thermocouples to measure the come-up time inside capillary tubes. Figure 2.1 shows unpublished data from the author's own experiments. Failure to account for come-up time will result in a higher survival rate than expected and produce a thermal death time curve (*z* value) that is concave upward at high temperatures. The larger the tube the longer the come-up time. Food slurries generally have a slower heat penetration rate than buffer or water. It is often advantageous to heat a large volume of food slurry to the desired temperature and then inject a small volume of suspension of micro-organism while mixing the food to aid temperature distribution. The resultant temperature drop of product/ suspension mixture and come-up time can be altered by altering the ratio of volume of product to volume of suspension of micro-organisms.

Steam is a better heating medium than water or oil because of the latent heat energy which is released when steam condenses. Steam heated systems like the thermoresistometer [35] can give extremely rapid sample come-up times provided that the sample is small and in a thin layer. Product into steam methods such as MISTRESS also utilise the latent heat energy of steam to effect rapid sample temperature rise. Droplet size in the MISTRESS is not controlled and it is possible that droplets coalesce and these larger droplets will heat up more slowly. This may account for some of the thermal death time curves obtained by Neaves and Jarvis [13, 14] for spores of *C. botulinum* heated in this system. In the author's own experience, it has been found that excessive condensate in the steam in the thermoresistometer can slow down the heating rate. Droplets of condensate which build up on the thin wire thermocouples in the thermoresistometer have also produced spurious results. It cannot be stressed too highly that working at the extremes of any method requires caution in the interpretation of results obtained. The micro-organisms will reflect the heat treatment they have experienced; the difficulty often lies in determining exactly what the heat treatment was.

Continuous flow systems involve an additional factor, residence time. Micro-organisms that travel fastest through a flow-through heating system will be more likely to survive than those which have a longer residence time [56]. The problem with these methods is determining both the residence time and the rate of heating. Residence time is often measured by injecting salt or dye and measuring the time for it to appear at the end of the heating or holding section. It is usually assumed that the micro-organisms will behave similarly but this may not be the case. Heppell [57] demonstrated the effect of changing from water to milk in a UHT plant and the effect on spore survival. Milk had a shorter mean residence time than water and this resulted in higher survivor levels in milk compared to water.

It is customary to use thermocouples to measure the temperatures achieved during heat resistance experiments. Smaller thermocouples have a more rapid response to temperature changes. However, the more sensitive the thermo-couple, the more likely it is to be affected by small temperature changes. This can be quite disconcerting to the investigator who may be expecting a smooth temperature curve only to find that every small temperature fluctuation is monitored by a thin wire thermocouple. Larger thermocouples and thermometers will dampen down the measured fluctuations and may falsely indicate smooth temperature profiles when this is not the case. Another problem in temperature measurement may be conduction of heat along the thermocouple wire to the measuring point. This is usually more of a problem with thick wire thermocouples or where temperature gradients are large. The response time of the thermocouple and the measuring equipment should be compatible. If the thermocouple responds much faster than the measuring equipment then temperature readings will lag behind actual temperatures. If the thermocouple is more sluggish than the recorder, a similar effect will be observed but in this case it will be the thermocouple that will be the problem.

Thermocouples should be calibrated against a recognised reference instrument at several temperatures within the desired experimental range. Thermocouple outputs do not usually agree with published tables in practice. Regular calibration is necessary since response may change with age. Pflug [58] details many of the precautions necessary when using and calibrating thermocouples used in heat resistance studies.

Time of exposure is best measured automatically, if possible, for short exposure times. In the thermoresistometer built by Brown *et al.* [43] exposure times were cross-checked using electronic timers and a computer clock which were triggered by microswitches. The recorded running time was also logged automatically alongside temperatures to give a complete time–temperature profile for each experiment. It should be noted that no matter how quickly a sample is transferred to the heating device (e.g. capillary tube into an oil bath) the rate of heating will depend on the rate of heat transfer which is in turn dependent on the thermal diffusivity of the sample, the size of sample, the surface area in relation to volume and the energy available in the heat source. There is no such thing as instantaneous heating; simply that some methods heat faster than others.

2.4 Handling survivor data

Schmidt [59] stated that the only practical criterion of the death of a micro-organism is its failure to reproduce when favourable conditions are provided. However, this definition raises a number of problems in determining 'favourable' conditions. Addition of soluble starch or lysozyme to recovery media for recovery of heat stressed bacterial spores is known to increase survivor counts. Thus a 'survivor' on one medium may be classed as 'dead' on another less favourable recovery medium. In spite of such difficulties, the most common methods of enumerating survivors of a heat treatment depend on the ability of the organism to grow in a particular recovery medium.

There are two main methods of enumerating survivors. The first is to culture the survivors on agar which has been supplemented to provide suitable recovery conditions. In a typical series of experiments, a number of replicate samples would be heated for different time intervals at a fixed temperature. The samples would then be diluted and aliquots plated out. After appropriate incubation conditions, the numbers of colonies which had developed would be counted. The results would then be plotted as the logarithm of the number of survivors per unit against process time (Figure 2.2).

The second method, called the fraction negative method, is used for gathering data for low survivor levels. Several replicates are heated for different times at a fixed temperature and survival estimated as a presence/absence (growth/no growth) procedure. A similar procedure which was

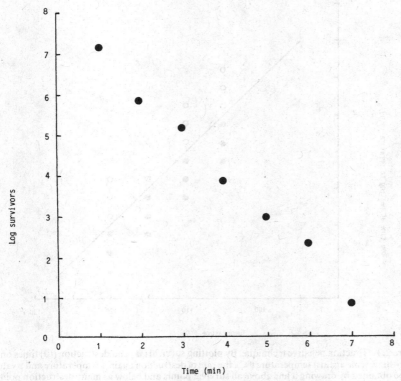

Figure 2.2 Plot of log survivors against time (min).

common in the 1930s was to heat replicates at different temperatures to obtain survival/destruction data. By plotting the log of the survival/destruction times against temperature, the change in the rate of destruction with temperature could be evaluated (Figure 2.3).

It is usually assumed that thermal death proceeds at an exponential rate such that by plotting log survivors against time a straight line is obtained. This is not always the case and a variety of different shapes for survivor curves have been reported (Figure 2.4). In some cases a deviation from logarithmic kinetics can be explained, for example, by faults in the experimental procedure, activation of spores, clumping of micro-organisms or heating of mixed cultures.

2.5 Process evaluation

There are a number of detailed articles dealing with the mathematical treatment of survivor data and its relationship with thermal process

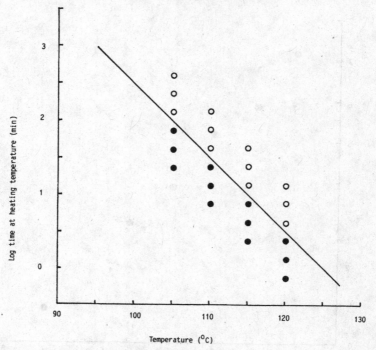

Figure 2.3 Fraction negative technique. By plotting survival (●) and destruction (○) times on a logarithmic scale against temperature (°C), the rate of destruction against temperature and z value can be obtained by drawing a line above all survival points and below as many destruction points as possible.

calculations. The reader is referred to Stumbo [5] and Pflug [58] for detailed accounts. A simplified introduction is presented here.

2.5.1 Decimal reduction time or D value

The decimal reduction time or D value is the time at any temperature to destroy 90% of the spores or vegetative cells of a given organism. It is equal to the number of minutes for the survivor curve to traverse one log cycle and can be calculated from the reciprocal of the slope of the survivor curve (Figure 2.5). It is related to the number of survivors as follows: assuming a logarithmic death rate

$$N = N_0 e^{-kt} \tag{2.1}$$

where N = final number of survivors after heat treatment, N_0 = initial number of organisms, e = exponential function, k = thermal death rate coefficient

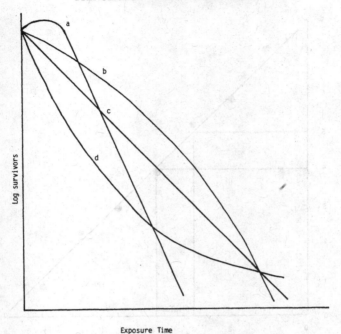

Figure 2.4 Examples of survivor curves deviating from logarithmic destruction. (a) Activation shoulder; (b) concave down; (c) logarithmic; (d) tailing.

(traditionally measured in s^{-1}) and $t =$ time. Rearranging gives

$$\log N = \log N_0 - \frac{60kt}{2.3} \tag{2.2}$$

(plotting $\log N$ against t in minutes gives the slope $-60k/2.3\,\text{min}^{-1}$). The inverse of the slope is

$$D = 2.3/60k\ \text{min} \tag{2.3}$$

The D value may be calculated from the inverse of the slope of the survivor curve by regression analysis of the data points where slope is given by

$$\text{Slope} = \frac{(\sum U_i(\log N_i) - [(\sum U_i)(\sum \log N_i)]/n)}{\sum U_i^2 - [(\sum U_i)^2/n)]} \tag{2.4}$$

where $U =$ equivalent heating time at heating medium temperature, and i denotes the ith observation of the n individuals $(U, \log N)$ [58]. Such calculations can easily be handled by pocket scientific calculators. D values can also be obtained from fraction negative data using

$$D = \frac{U}{\log N_0 - \log N_u} \tag{2.5}$$

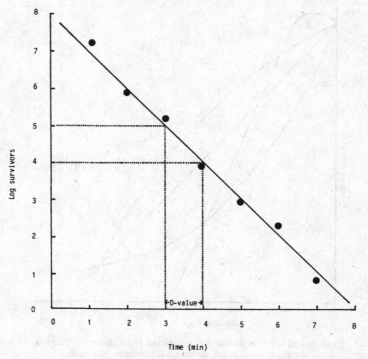

Figure 2.5 Plotting log survivors against time (min) at a particular temperature (e.g. 121.1°C) in order to obtain the D value.

where N_u = number of survivors after a process of time u at the heating medium temperature.

If a presence/absence (growth/no growth) method is used to estimate the number of survivors then N_u can be calculated using the Halvorson and Ziegler equation

$$N_u = 2.303 \log(n/r) \tag{2.6}$$

where n = number of units heated, r = number of units sterile and N_u = most probable number of survivors.

2.5.2 Thermal death time constant or z value

The change in D value with temperature can be obtained by plotting log D against temperature (the thermal resistance curve, Figure 2.6). The z value is the number of degrees for the thermal resistance curve to traverse one log cycle and is equal to the reciprocal of the slope of the curve. The equation applied to

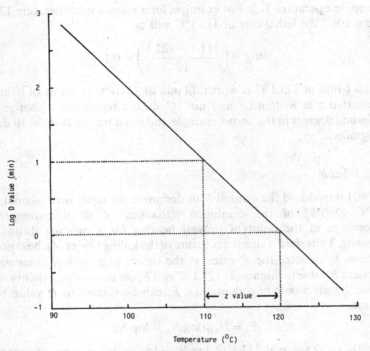

Figure 2.6 Plotting log D value (min) against temperature (°C) in order to obtain the z value.

the thermal resistance curve is given by [5]

$$\log D_{Ref} = \log D_T = \frac{1}{z(T - T_{Ref})} \tag{2.7}$$

where $D_{Ref} = D$ value at the reference temperature, $T_{Ref} = 121.1°C$, $D_T = D$ value at another temperature T and z = thermal death time constant.

Originally the z value was obtained by plotting survival/destruction data on a logarithmic scale against temperature [6] (Figure 2.3). The thermal death time curve was drawn above all survival points and below as many destruction points as possible.

2.5.3 Lethal rate

Using the z value, the lethal rate L can be calculated from [5]

$$L = \log^{-1}\left(\frac{T - T_{Ref}}{z}\right) \tag{2.8}$$

The lethal rate is a measure of the lethality of any temperature T relative to the reference temperature T_{Ref}. For example, for a reference temperature 121.1°C and $z = 10C°$ the lethal rate at 111.1°C will be

$$\log^{-1}\left(\frac{111.1 - 121.1}{10}\right) = 0.1$$

Thus 1 min at 111.1°C is worth 0.1 min at 121.1°C in terms of lethality.

Note that z is written $C°$ and not $°C$ since z represents a change of 10 Centigrade degrees in the above example and not a temperature of 10 degrees Centigrade.

2.5.4 F value

Ball [60] introduced the symbol F to designate the equivalent in minutes at 121.1°C (250°F) of the combined lethalities of all time–temperature relationships at the point of slowest heating for a product during heat processing. Thus the F value is a measure of the killing power of a heat process. The term F_c denotes the F value at the centre of a pack, F_0 denotes the equivalent F value in minutes at 121.1°C and F_s the integrated lethality of heat received by all points in a container. F_s can be related to D value by the equation

$$F_s = D_{Ref}(\log N_0 - \log N) \tag{2.9}$$

where $D_{Ref} = D$ value at 121.1°C, $\log N_0 = \log$ (initial number of organisms) and $\log N = \log$ (final number of organisms). In Eq. (2.9), F_s can be considered as equivalent to F_0 or F_c in rapid heating systems such as the thermoresistometer. The F_0 value of a process can in practice be obtained by summing the lethal rates at 1 minute intervals from the heating and cooling curve of a product during a heat process. Note that this is a simplified view and process calculations in the industry involve rather more complicated mathematics and usually do not include the cooling curve contribution to the process. The reader is referred to the worked examples in Stumbo [5] which show in detail how to calculate process values using the basic equation

$$B = f_h(\log J_{ih}I_h - \log g_c) \tag{2.10}$$

where B = thermal process time corrected for time to bring retort to process temperature, f_h = time in minutes for semi log heating curve to traverse one log cycle, j_{ih} = heating lag factor, I_h = difference in temperature between retort and food at start of process and g_c = difference in temperature between retort temperature and the maximum temperature reached by the food at the centre.

In the processing of low acid foods with a pH value of 4.5 or above, a process equivalent in lethality to at least $F_0 = 3$ min must be applied to minimise the risk of spores of C. botulinum surviving the process. It is customary to use a z value of 10C° in the case of C. botulinum. In order to convert an F value determined on the basis of one z value to an F value determined on the basis of a second z value the reader is referred to an article by Pflug and Christensen [61].

2.5.5 Cook value or C value

Mansfield [62] introduced the concept of a lethality-like value for sensory degradation termed the cook value or C value. This has a reference temperature of 100°C and z value typically in the range 20–40C° [63].

$$C_{100} = 10^{(T - 100)/z} c \text{ min}$$

2.5.6 Pasteurisation units or PU value

At temperatures below 100°C it is most convenient to use pasteurisation or PU values instead of F values. PU values are calculated in much the same way as F values using

$$PU = \log^{-1}\left(\frac{T - T_{Ref}}{z}\right)$$

The main difference with F values is that whereas F values usually have a standard reference temperature of 121.1°C and $z = 10C°$, PU values do not have a standard reference temperature or z value. Instead, processors use a reference temperature that is appropriate to their particular process, e.g. 70°C, 82.2°C (180°F) or 90°C, and a z value appropriate for the organisms they wish to control. This is a source of considerable confusion to the layman who, not unreasonably, assumes that pasteurisation values have a common reference point. It is very important that the reference temperature and z value of PU values are always specified. PU is also found in a shortened form, P value. A P value P_{65}^5 means a P value calculated with a reference temperature 65°C and $z = 5C°$. For further information on P values the reader is referred to Shapton [64] and Shapton et al. [65]. At the time of writing, a document on 'Guidelines to Food Products Stabilised by Pasteurisation Treatments' is being compiled by members of a pasteurisation working party at the Campden Food and Drink Research Association.

2.5.7 The botulinum cook

The spores of $C. botulinum$ are sufficiently heat resistant to survive processing above 100°C. It was this property which resulted in the development of the minimum botulinum cook, which is a process equivalent in lethality to 3 min at 121°C ($F_0 3$) based on a thermal death coefficient or z value of 10C°. The use of this concept has proved for over 50 years to be adequate for the calculation of processes for low acid heat processed foods (pH 4.5 and above) to ensure that they are safe from the risk of survival of spores of *C. botulinum*. There have been no outbreaks of botulism due to spores of *C. botulinum* surviving an $F_0 3$ process.

A botulinum cook is often referred to as a 12-D process. However a 12-D process and a process equivalent in lethality to 3 min at 121°C ($F_0 3$) are not necessarily the same. The process $F_0 = 3$ is calculated completely independently of the 12-D concept. The 12-D concept is based on several assumptions including the heat resistance, the distribution and concentration of the spores, assumptions which are unlikely to hold true in practice. It is also important to realise that assumptions regarding the kinetics of death for 12 log reductions are often based on experiments with only perhaps 7 log reductions being achieved (i.e. extrapolation over 5 log reductions). In terms of probability of survival, the 12-D concept equates to a probability of 1×10^{-12}. The 12-D concept is several orders of magnitude beyond current experimental practices used to determine resistance of the spores of *C. botulinum*. Fortunately the botulinum cook is not based on such vanishingly small probability estimates but on proven safety in practice. It should also be noted that the probability of survival per can will be different to the probability of survival per batch. If the probability of survival of *C. botulinum* spores is one in 10^{12} containers, in batches of 100,000 containers the probability per batch will be 1 in 10^7.

2.6 Recommended and statutory processes

2.6.1 Low acid foods ($pH \geqslant 4.5$)

The heat processing of low acid ($pH \geqslant 4.5$) foods in the United Kingdom is covered by the Food Hygiene Code of Practice No. 10 [66]. The minimum thermal process for a low acid canned food must reduce the probability of survival of spores of *C. botulinum* to less than 1 in 10^{12} containers. This is usually interpreted as a minimum thermal process value $F_0 = 3$. In practice, processors usually apply heat processes in excess of $F_0 3$ (e.g. 6–7 or more) to ensure control of spoilage organisms. Code of Practice No. 10 is currently under revision and the revised version should be available in late 1990 or early 1991.

The readers should also refer to the Food Processors Institute (USA) book on canned foods [67], Hersom and Hulland [1], Pflug [58], Campden Food Preservation Research Association Technical Manual No. 3 [68] and Technical Bulletin No. 4 [69], Codex Alimentarius [70], The National Canners Association (USA) Laboratory Manual for Food Canners and Processors, Vol. 1 [71], National Canners Association (USA) Bulletin 26-L [72] (Processes for low acid canned foods in metal containers) and Bulletin 30-L [73] (Processes for low acid canned foods in glass containers), Lopez [74] and the Food and Drug Administration (USA) Federal Regulations, parts 113 and 114 [75, 76].

These various manuscripts provide guidance on recommended process times and temperatures for different products and container sizes. It cannot be stressed too highly that it is good practice to confirm processes experimentally and also make allowances for production variation during the establishing of a thermal process. It is also recommended that processes are checked after any changes in production that may affect the thermal process.

2.6.2 Dairy products

In the United Kingdom, the heat processes for milk, cream, milk-based drinks and ice cream are subject to Statutory Instruments under the Food and Drug Regulations. A summary of the heat treatments for milk, cream and milk-based drinks taken from the regulations is given below.

(1) The Milk (Special Designation) Regulations 1977 (S.I. 1977 No. 1033; 16 June 1977) [77]

(a) Pasteurisation

- The milk shall be retained at a temperature of not less than 62.8°C and not more than 65.6°C for at least 30 min and be immediately cooled to a temperature of not more than 10°C; or
- retained at a temperature of not less than 71.1°C for at least 15 s and be immediately cooled to a temperature of not more than 10°C; or
- retained at such temperature for such period as may be specified by the licensing authority with the approval of the minister.

(b) Sterilisation

- The milk should be maintained at a temperature above 100°C for such a period that it complies with the turbidity test.

(c) Ultra heat treated

- The milk should be treated at a temperature of not less than 132.2°C for not less than 1 s.

(2) The Milk and Dairies (Heat Treatment of Cream) Regulations 1983 (S.I. No. 1509) [78]

(a) Pasteurisation

The cream shall be heated
- to a temperature of not less than 63°C and retained at that temperature for not less than 30 min; or
- to a temperature of not less than 72°C and retained at that temperature for not less than 15 s; or
- to such other temperature or period that is equivalent in lethality.

(b) Sterilisation

The cream shall be heated
- to a temperature of not less than 108°C and retained at that temperature for not less than 45 min; or
- to such other temperature or period that is equivalent in lethality.

(c) Ultra heat treated

The cream shall be heated
- to a temperature of not less than 140°C and retained at that temperature for at least 2 s; or
- to such other temperature or period that is equivalent in lethality.

(3) The Milk-Based Drinks (Hygiene and Heat Treatment) Regulations 1983 (S.I. No. 1508) [79]

The regulations for the heat treatment of milk-based drinks are similar to cream with respect to the actual times and temperatures of processing.

The International Dairy Federation is currently revising its definitions of pasteurisation and sterilisation and gives examples of temperature/time combinations for pasteurisation of

- Pasteurised milk and skimmed milk   63°C for 30 min
                                       72°C for 15 s
- Pasteurised cream (18% fat)   75°C for 15 s
           (≥ 35% fat)   80°C for 15 s
- Pasteurised concentrated milk   80°C for 25 s

This organisation also refers to the sterilisation of milk as a process equivalent in lethality to a minimum $F_0 = 3$. At 140°C this is equivalent to 2.3 s. It should be noted that an $F_0 = 3$ process may not be sufficient to destroy spores of spoilage organisms, e.g. thermophilic bacteria.

The IDF process times and temperatures are similar to the figures given in the FDA (USA) Recommended Guidelines for Controlling Environmental Contamination in Dairy Plants [80]. These figures and the UK definition of milk pasteurisation (72°C for 15 s) are also cited in the Guidelines for Good Hygienic Practice in the Manufacture of Soft and Fresh Cheeses issued by the Creamery Proprietors Association [81].

2.6.3 *Acid products (pH < 4.5)*

At pH values below 4.5 the risk of growth and toxin production by *C. botulinum* is extremely unlikely and for products with pH values between 4.0 and 4.5, processes are aimed at controlling the survival and growth of spore-forming organisms such as *Bacillus coagulans, B. polymyxa, B. macerans* and the butyric anaerobes such as *Clostridium butyricum* and *C. pasteurianum*. Hersom and Hulland [1] regard a heat process of $F_{121}^{10} = 0.7$ as adequate for this purpose. In the Laboratory Manual for Food Canners and Processors, Vol. 1 [71] the National Canners Association (now the National Food Processors Association) of the USA recommended a process equivalent to 10 min at 93.3°C ($F_{93.3}^{8.3} = 10$) when the pH is between 4.3 and 4.5 and a process equivalent to 5 min at 93.3°C ($F_{93.3}^{8.3} = 5$) when the pH is between 4.0 and 4.3. They warn, however, that more severe processes may be required to control excessive contamination. Hersom and Hulland [1] cite work by Slocum *et al.* [82] who investigated nine cases of botulism from products with pH below 4.5 including canned pears, canned apricots, tomato catsup, tomato-onion chilli sauce and green tomatoes. It was suggested that growth of spoilage organisms raised the pH to a level at which dormant spores of *C. botulinum* could germinate and grow.

Below pH 3.7 the processor is concerned with the control of non-sporing bacteria, yeasts and moulds. These may be generally controlled by heat processes at temperatures below 100°C. Attention is drawn, however, to the sections in this chapter on the heat resistance of viruses and the moulds *Byssochlamys fulva* and *B. nivea*.

2.7 Heat resistance of micro-organisms

2.7.1 *Bacterial spores*

The most important bacterial spore-former with respect to heat processing is *Clostridium botulinum* because of the potent neurotoxin that it produces. The organism occurs in seven distinct serotypes A–G which are further subdivided into proteolytic and non-proteolytic strains. The most heat resistant spores are those produced by type A and the proteolytic B strains. The D value at 121.1°C of the most resistant strains is generally considered to be 0.21 min [1]. Typical heat resistance values for the different strains of *C. botulinum* responsible for human food poisoning are presented in Table 2.3.

Table 2.3 Heat resistance of spores of *C. botulinum*.

Strain	Temperature (°C)	Typical D value (min)	Typical z value (C°)	Reference
Type A (proteolytic)	121.1	0.13	9.0	[83]
Type B (proteolytic)	121.1	0.15	11.0	[84]
Type B (non-proteolytic)	82.2	1.5–32.3	8.3–16.5	[85]
Type E (non-proteolytic)	77.0	0.77–1.95	—	[86]
Type F (proteolytic)	121.1	0.14–0.22	9.3–12.1	[87]
Type F (non-proteolytic)	77.0	1.6–9.5	—	[88]
Type G (proteolytic)	115.6	0.25–0.29	20.9–27.3	[89]

The proteolytic strains of *C. botulinum* typically produce gas and a putrefactive odour during growth in food whereas the non-proteolytic strains may produce little organoleptic change in the product.

Pflug and Odlaug [90] recently reviewed the published data on heat resistance up to 126.7°C for the proteolytic strains of *C. botulinum* and recently Gaze and Brown [84] published further data up to 140°C to cover the temperatures used in high temperature (UHT) processing of foods. As can be seen in Figure 2.7 the data from both Pflug and Odlaug [90] and Gaze and Brown [84] are in agreement with the classical Esty and Meyer [91] data indicating that the use of a single z value of 10C° can be used for calculation of process values up to 140°C. This supports the view that the traditional botulinum process ($F_0 = 3$) can be safely extrapolated into the UHT region in order to calculate thermal process values for UHT processed products.

Outbreaks of botulism from canned foods have often been associated with home canning. In the United States where home canning is common, there were 688 reported outbreaks of botulism between 1899 and 1973 resulting in 978 deaths [92] of which 72% were due to home canning.

In a recent unusual outbreak of botulism [93] involving fried onions, spores of *C. botulinum* appear to have survived the frying process and then grown in the anaerobic environment provided by the margarine. Under normal frying conditions the temperature inside a particle of moist food does not exceed 100°C even though the temperature of the fat or oil used for cooking may be at temperatures of 165–185°C [94]. Two other outbreaks of botulism have been reported in California which more closely resemble outbreaks of *Clostridium perfringens* in their cause. In both outbreaks, a stew or meatloaf was left in a warm oven for between 16 and 24 h after cooking which allowed the spores to germinate and grow and produce toxin. In both cases, cooking temperatures in the product would not have exceeded 100°C.

In 1989 there was a major outbreak of botulism associated with hazelnut yoghurt in the United Kingdom in which 27 people developed botulism and one died. This was caused by failure to apply a botulinum cook to the canned hazelnut used to flavour the yoghurt. This example of botulism from a product with pH below 4.5 emphasises the need for vigilance throughout a production process from raw materials to finished product, including any manufacturing stages en route.

Clostridium sporogenes, which is closely related to the proteolytic strains of *C. botulinum*, produces spores which are more heat resistant [95, 96] (Table 2.4). The *D*-value of the spores at 121.1°C can be up to 1.5 min. Hersom and Hulland [1] reported that *C. sporogenes* was the most common mesophilic putrefactive anaerobe in the spoilage of low acid (pH ≥ 4.5) canned foods. Matsuda *et al.* [99] identified over 53% of the isolates they examined from canned foods as *C. sporogenes*.

Clostridium perfringens was the cause of 18 970 cases of food poisoning between 1970 and 1980 in England and Wales [100]. The symptoms are

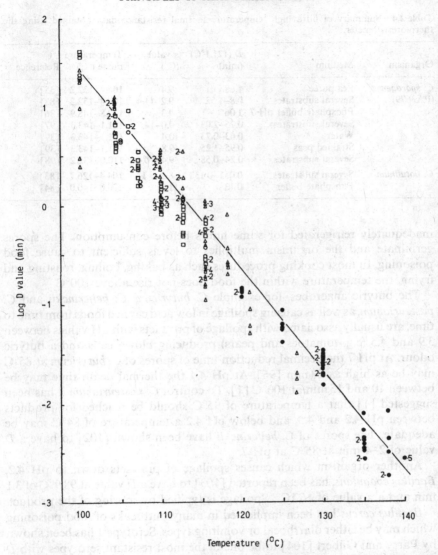

Figure 2.7 Comparative heat resistance data for *C. botulinum* over the temperature range 100–140°C; Pflug and Odlaug [90] 213B buffer (○), 213B product (□), other strains (△); Gaze and Brown [84] (●). The line indicates the classical Esty and Meyer [91] line.

typically diarrhoea and abdominal pains 8–24 h after ingestion of food containing high numbers of vegetative cells. The spores of *C. perfringens* may survive from a few minutes to hours at 100°C depending on strain and substrate [101]. In a typical outbreak of food poisoning, the spores survive the cooking process and the food is then left at room temperature or is

Table 2.4 Summary of ultra high temperature thermal resistance data obtained using the thermoresistometer.

Organism	Medium	D (121.1°C) (min)	z value (C°)	Temperature range (°C)	Reference
C. sporogenes	Pea purée	1.68	9.4	104.4–132.2	[35]
(PA3679)	Several substrates	0.8–1.52	9.2–11.4	104.4–132.2	[83]
	Phosphate buffer pH 7	1.06	9.3	112.8–148.9	[36]
	Several substrates	0.75–2.03	9.0–14.7	121.1–143.3	[97]
	Water	0.63–0.73	10.4	121.1–143.3	[39]
	Strained peas	0.95–1.25	9.8	121.1–143.3	[39]
	Several substrates	0.24–0.58	9.4–10.4	110.0–132.2	[98]
C. botulinum	Several substrates	0.051–0.133	8.2–9.1	104.4–126.7	[83]
	Phosphate buffer	0.13	11.0	120.0–140.0	[84]

inadequately refrigerated for some hours before consumption. The spores germinate and the organism multiplies to levels sufficient to cause food poisoning. In most cooking procedures such as baking, boiling, roasting and frying, the temperature within the food does not rise above 100°C.

The butyric anaerobes, for example *C. butyricum, C. beijerinckii* and *C. pasteurianum,* as well as causing spoilage in low acid canned foods from time to time, are usually associated with spoilage of products with pH values between 3.9 and 4.5 (e.g. tomatoes and pears) producing blown cans and a butyric odour. At pH 7 the decimal reduction time of spores of *C. butyricum* at 85°C may be as high as 23 min [96]. At pH 4.4 the thermal death time may be between 10 and 15 min at 100°C [1]. To control *C. pasteurianum* it has been suggested [1] that a temperature of 95°C should be reached for products between pH 4.2 and 4.5, and below pH 4.2 a temperature of 84°C may be adequate. The spores of *C. beijerinckii* have been shown [102] to have a D value of 2–4 min at 85°C at pH 7.

Another organism which causes spoilage of products down to pH 4.2, *Bacillus coagulans,* has been reported [103] to have a D value at 98.9°C of 3.1 min and a z value of 16.1C°. Spoilage is typified by souring of the product.

Bacillus cereus has been implicated in many outbreaks of food poisoning which may be either diarrhoeal or vomiting types. Serotype 1 has been shown by Parry and Gilbert [104] to be one of the most resistant serotypes with D values at 95°C ranging from 22.4 to 36.2 min. *B. cereus* food poisoning, particularly the vomiting type which has an incubation period of 1–5 h, has often been associated with Chinese cooked rice dishes.

Other mesophilic *Bacilli* which produce heat resistant spores and have been implicated in food poisoning outbreaks are *B. subtilis* and *B. licheniformis.* Typical D values for these organisms are shown in Table 2.5.

Spores of the thermophilic organisms which have an optimum growth temperature around 55°C are usually much more resistant to wet heat than the mesophilic. The heat resistance of the most resistant thermophiles is given in

Table 2.5 Typical heat resistance of selected mesophilic sporeformers [96].

Organism	Temperature (°C)	Typical D value (min)
Bacillus cereus	100	5.5
	121	2.37
Bacillus coagulans	121	3.0
Bacillus licheniformis	100	13.0
Bacillus subtilis	121	0.3–0.7
Bacillus coagulans	96	8.0
Clostridium butyricum	85	12–23
Clostridium sporogenes	121	0.2–1.5
Clostridium perfringens	100	0.3–17.6

Table 2.6 Heat resistance of spores of some particularly heat resistant thermophilic spore-formers.

Organism	D (121°C) (min)	z (C°)	Reference
Bacillus stearothermophilus	16.0	7.7	[20]
Desulfotomaculum nigrificans	55.0	9.5	[105]
Clostridium thermosaccharolyticum	68.0	11.0	[106]
	195.0	6.9	[107]

Table 2.6. The most resistant thermophile is *Clostridium thermosaccharolyticum* which produces blown cans and a butyric or 'cheesy' odour, with D values as high as 68–195 min at 121.1°C. Segner [108] reported 14% of spoilage outbreaks investigated by the Continental Can Company were caused by thermophiles. Spoilage from *Desulfotomaculum nigrificans*, which causes 'sulphur stinker' spoilage is rare but the spores may have D values as high as 54.4 min at 121.1°C. The spores of *Bacillus stearothermophilus* have often been used in process evaluation studies because of their high heat resistance. The organism can cause spoilage of products with pH values above 5.3. Spoilage results in souring of the product with no gas production. Thermophilic spoilage in canned foods usually occurs when the spores have survived the heat process and the cans are either inadequately cooled or are exported to hot climates.

Matsuda *et al.* [109] reported an interesting outbreak of thermophilic spoilage in hot drink vending machines in Japan caused by *D. nigrificans* and *Clostridium thermoaceticum* which had D values at 121.1°C of 2.6 and 44.4 min, respectively.

Another unusual bacterial sporeformer is *Sporolactobacillus inulinus*. This organism has been isolated from a number of food and environmental sources and has been reported [110] to have a D value at 90°C of 5.1 min and a z value of 13C°. Quality control personnel should therefore be aware that not all isolates of gram-positive spore-forming rods are automatically of the genera *Bacillus* or *Clostridium*.

Table 2.7   Heat resistance of some vegetative bacteria.

Organism	Heating medium	D value (min) at 70°C	z (C°)	Reference
Escherichia coli	Nutrient broth	0.006[a]	4.9	[111]
	Milk	0.04[a]	6.5	[111]
Lactobacillus casei	Tomato juice	4.0	11.5	[111]
L. plantarum	Tomato juice	11.0	12.5	[111]
Listeria monocytogenes	Raw beef	0.15	6.7[a]	[112]
L. monocytogenes	Chicken, beef and carrot homogenates	0.14–0.27	5.98–7.39	[113]
Salmonella typhimurium	Aqueous sucrose/ glucose (A_w 0.995)	0.03[a]	17.05	[111]
Salmonella typhimurium	Milk chocolate	816	19.0	[111]
Salmonella typhimurium	51% milk	0.35[a]	6.8	[111]
Staphylococcus aureus	Milk	0.30[a]	5.1	[111]
Streptococcus faecalis var. zymogenes	Broth	2.84[a]	17.0	[111]
Strep. faecium	Broth	0.015[a]	3.5	[111]
Strep. faecium	Ham	2.57[b]	6.8–7.5	[114]
Microbacterium lacticum	Skim milk	4.0		[111]

[a]Calculated from information presented in the reference.
[b]D value at 74°C.

2.7.2   Vegetative bacteria

In comparison with bacterial spores, vegetative bacteria are not very heat resistant under normal wet heating conditions (Table 2.7). When heated in fat or 'dry' situations (e.g. particles of dried powder) then there is evidence that vegetative bacteria (and spores) can be much more resistant. The literature is reviewed by Hersom and Hulland [1] who give examples of streptococci requiring 20 min at 120°C in dry butter for destruction and spores of *Bacilli* surviving 100 to 1000-fold longer also in butter. *Salmonella typhimurium* may have a D value of over 800 min at 70°C in chocolate (Table 2.7). Thus while the figures for heat resistance of vegetative bacteria given in Table 2.7 can be used as a guide, it should be appreciated that the characteristics of the food product can have a pronounced effect on survival of micro-organisms.

2.7.3   Yeasts

Yeasts are rarely involved in spoilage of heat preserved foods except in the case of gross underprocessing or leakage. They are usually associated with spoilage of acid products, fruits and jams [1]. In the author's experience, leaker spoilage involving yeasts is usually associated with obvious container defects. Yeasts can also affect low acid products, for example an outbreak of leaker spoilage in canned carrots caused by yeasts investigated by the author (unpublished).

Put and de Jong [115] reported D values for ascospores of *Saccharomyces* species of 1.0–22.5 min with z values between 4.0 and 6.5C° and D_{60} values for *Kluyveromyces bulgaricus* between 20 and 40 min.

2.7.4 Moulds

By far the most common spoilage moulds in heat preserved acid products such as strawberries and grapes are the species *Byssochlamys fulva* and *B. nivea*. Spoilage is manifested by disintegration of the fruit often with little evidence of visible fungal mycelium. It has been reported [1] that growth can take place in a vacuum of 50.8 cmHg. Heat processed strawberries are frequently associated with spoilage by *B. fulva* and *B. nivea* for a number of reasons. The moulds are soil borne, strawberries are easily contaminated by soil and during processing under static conditions, the fruit clumps together and heat transfer is reduced, leading to process survival. The author has also received reports of *Byssochlamys* spoilage in strawberry pulp processed in a tubular pasteuriser and spoilage due to inadequate sterilisation of bulk containers used for temporary storage of aseptically filled processed strawberry pulp used for yoghurt manufacture.

2.7.5 Bovine spongiform encephalopathy

Bovine spongiform encephalopathy or BSE is a degenerative brain disease of cattle, closely related to a similar disease called scrapie in sheep and Creutzfeldt-Jacob disease in humans. BSE hit the headlines in 1989 because it was feared that the disease could be spread from affected cattle to man via offal and meat pies. In November 1989, the use of certain specified bovine offal was banned for human consumption in the United Kingdom because of the possible spread of BSE to man. This followed a ban in February 1989 on the use of bovine offal in baby food, although few if any, baby food manufacturers were actually using bovine offal. The organism which causes scrapie and BSE is very heat resistant and normal cooking temperatures are unlikely to destroy it. The 1984 DHSS guidelines for autoclaving of materials contaminated with Creutzfeld-Jacob disease were either [116]

- a single cycle at 134°C ($\pm$4) (30 psi) for 18 min holding time at temperature; or
- six separate cycles at 134°C ($\pm$4) (30 psi) for 3 min holding time at temperature.

These cycles are, of course, very much more severe than those processes usually employed in the heat preservation of meat products. The best precaution therefore is to exclude meat potentially at risk.

In the UK the following statuatory regulations have been introduced;

(i) All carcasses from infected animals to be destroyed.
(ii) Ruminants cannot be fed diets containing ruminant derived protein.

(iii) All cattle brain, spinal cord, spleen, thymus and tonsil from healthy animals are banned from human consumption.

Manufacturers of human and pet foods should follow these precautions stringently and conscientiously.

2.7.6   Heat resistance of viruses in food

Occasionally outbreaks of viral food poisoning result from inadequately processed foods. An example of a product group that gives rise to many cases of viral food poisoning is molluscan shellfish. Guzewich and Morse [117] reported 1293 cases of hepatitis A from shellfish over the period 1961–1981 in the United States. Hepatitis A virus is inactivated in water after 5 min at 100°C. At 60°C for 1 h the titre is reduced. West et al. [118] suggested that a temperature of 98°C for 1 min is required to inactivate hepatitis A virus in shellfish. On this basis some of the pasteurisation processes used for pickled shellfish in vinegar may not be adequate to ensure freedom from viruses. Parvoviruses [119] have also been implicated in gastrointestinal infections associated with consumption of raw shellfish. Parvoviruses have unusually high thermal resistance with 50% reduction in .numbers (fraction negative = 50) after 0.2 min at 78°C for H-1 virus to 4.3 min at 90°C for rat virus. Polio virus is another heat resistant virus which may have a D value at 80°C of up to 2.3 min [120].

2.8   The statistics of sampling for detection of contamination

2.8.1   Contamination level and sampling considerations

Heat preserved products are intended to be commercially sterile. By design, no pack should be infected with viable micro-organisms capable of growing in the food under normal storage conditions. From time to time packs will be contaminated, and while zero contamination must remain the target, the manufacturer will have to choose an acceptable spoilage level at which action will be taken. It follows that decisions involving the release for sale of batches with more than zero infected units will have two essential requirements. The first is that there is no chance that the contaminants detected can be of any public health significance. The second is that any level of contamination must be very low.

Any sampling plan trying to detect very low contamination levels will inevitably require a high number of sample units. As an acceptable contamination level in this case is always below 1 in 1000 packs, it is certain that such sampling plans are not going to be applied routinely because of the prohibitively high number of sample units needing to be examined to be confident of identifying an unacceptable batch. For example, reference to Table 2.8 shows that with a critical defect level as high as 1 in 1000, at least

Table 2.8 Sampling plans for 0.1% critical defect standard. For N sample units, there is an X probability of batch acceptance when A packs are contaminated if there are Y critical defects.

Sample units	Positives acceptable	Probability of acceptance
1000	0	0.37
2000	0	0.14
2000	1	0.59
3000	0	0.05
3000	1	0.20
3000	2	0.42
4000	0	0.02
4000	1	0.09
4000	2	0.24

3000 packs must be examined to be 95% sure of identifying such a batch as unacceptable.

Sampling will therefore not be used as an acceptance criterion for routine production; sample failures, however, will be clear evidence of a problem.

2.8.2 *Sampling philosophy*

Most statistically based sampling plans are based upon random sampling of a batch. In the case of aseptically packed foods this may be a very useful investigational tool. In most instances, random sampling is not likely to be appropriate.

One option is to use a very high sampling level, sometimes approaching 100%. This can be particularly useful when setting up or proving a process. In this case a high degree of confidence is required that the manufacturing system is capable of achieving the very low infection level required, and the operational and control systems installed are satisfactory for routine production. This technique is of particular use during commissioning stages.

2.8.3 *Sampling performance*

In routine quality control, sampling is normally carried out during production, requiring some orderly, non-random system of sampling. The normal statistics of sampling can then no longer apply. Nevertheless, this is a very powerful system of sampling if its limitations are understood.

In many processes, it is well recognised that there are 'timed' events which affect contamination, e.g. problems during start up, build up in the plant with time. Taking samples at set times (either by the clock, by the number of units produced or by some event in the process such as breaks in production) can enable failures to be identified and possible sources of the failure to be pinpointed. It is important to recognise that the relatively low level of sampling only enables major failures to be identified. Absence of contamination on such a routine test does not provide evidence that production meets an acceptable microbiological standard; this depends upon evidence that the scheduled process for the manufacturing operation was under control throughout the entire production run.

2.8.4 Commissioning trials

It may be useful during commissioning stages to know that the process is capable of producing packs with a very low level of contamination. Many manufacturers of packaging systems already recognise the need to carry out such tests on newly installed equipment. The principle of such tests is to run the equipment under normal conditions and to carry out a high level sampling over a series of production runs (rarely less than 1000 packs per run). These packs are then subjected to full incubation and examination for contamination. In many instances the product filled may be modified to make recognition of contamination easier.

This high level sampling requires an agreed operating standard for judging the results, and the process either meets the standard set or it does not. It is important for such tests that the processing operations and controls are all fully operational to avoid costly (or dangerous) misinterpretation of the results.

High level, random batch sampling can also be used for investigational purposes if there is an identified microbiological problem with a batch of product. If this type of test is carried out, the confidence which can be expressed in a result must be understood before decisions to release product based upon these results are made. This is particularly important if there is already evidence from the process controls that good manufacturing procedures were not followed.

2.8.5 Routine quality control testing

Sampling can either be on a timed basis during production, or selection of numbered units (e.g. 4 packs per pallet). No statistical interpretation can be placed upon the results from such a test, but it has the following advantages:

- Major contamination problems can be identified
- With time, the performance of the manufacturing process asepsis can be measured
- 'Hotspots', i.e. times or process stages of higher infection, can be identified over a period of time.

It is important to respond to problems observed during this routine sampling. This may need random batch sampling as indicated above. Often it is more efficient to go back to the batch and carry out more intensive sequential sampling from the batch. This allows for patterns of infection to be identified, often leading to an understanding of its cause during production. It is important in this case to have all samples and pallets sequentially coded so that such retrospective sampling can be carried out. Hence, if the percentage of critical defects is 1 in 1000 (0.1%, see Table 2.8) and we examine 3000 packs, there is a probability of 0.2 (a 20% chance) that we would find zero or one contaminated packs and accept the batch (and an 80% chance we would find more than one contaminated pack, and fail the batch).

2.9 Incubation and sample handling

Properly designed incubation tests may provide valuable information about the acceptability of a food sterilising or aseptic filling process or the microbiological status of any quantity of production. Wherever possible however, the results of incubation tests should not be relied upon as the sole criterion.

Incubation of filled containers is used to encourage proliferation of any micro-organisms that might be present and capable of growth in the product. Thus after an appropriate incubation period it should be possible to determine whether such organisms are present in the pack by cultural examination of only a few grammes of product. Alternatively, microbial proliferation may cause sufficient change in the food product as to be readily detectable by physical examination of the filled container or the product. Incubation tests may therefore be considered under the two main headings of laboratory incubation and bulk incubation.

Laboratory incubation tests are usually small scale followed by destructive examination of the samples. Their statistical soundness varies considerably and because of this, it is essential that the purpose of both the test and its statistical basis is clearly understood before any decisions or predictions are made from the results.

Bulk incubation tests can usually only be practicably used where spoilage is evident by visual examination of the container or physical examination of the contents. Since it is usually not possible to examine each individual container at intervals during the incubation period, this type of test is only suitable for incubation temperatures below 35°C (see below).

It should be borne in mind that some consequences of incubation are as follows:

- Examination of filled product is delayed for days or maybe weeks depending upon temperature of incubation.
- Only those organisms capable of proliferation in the product at the incubation temperature(s) used will be detected.
- Incubation temperature can be a selective influence.
- Proliferation of one species or group of organisms may prevent outgrowth of others.

- Prolonged incubation can result in senescence or death of organisms resulting in difficulty or inability to recover viable organisms.
- Curtailed incubation may not allow sufficient time for repair and growth of damaged organisms.

The choice of incubation time and temperature therefore requires careful consideration and may vary depending upon the information sought, type of product examined and preservation process used.

Where incubation temperatures are in excess of 35°C, to avoid pack leakage, it is important that samples are inspected frequently during the incubation period. Samples showing physical evidence of spoilage are removed for immediate examination. This requirement may limit the number of samples that can be incubated at higher temperatures.

The following filled container incubation regimes of time and temperature are suggested to encourage growth of different groups of micro-organisms. Note that incubation times are frequently longer than those used for laboratory media since spoilage of product in finished packs may often take longer to develop.

2.9.1 Thermophilic micro-organisms

Incubation temperature should be in the range 45–65°C, the most commonly used temperature being 55°C. However, it has been suggested that outgrowth of those thermophiles also capable of growth below 40°C (e.g. Bacillus coagulans) can be selectively encouraged by incubation at 45°C.

Since products take some time to warm up to higher incubation temperatures, it is important to ensure that containers are sufficiently separated from each other so as to heat as rapidly as possible. If there is significant delay in reaching the incubation temperature, other non-thermophilic organisms may be able to grow and confuse the test.

For this reason it is often not practicable to carry out bulk incubation at higher temperatures with cold-filled aseptic products.

Incubation times of 5–14 days have been recommended for thermophilic organisms. Whatever time period is used it is important to recognise that some thermophilic spore-forming bacteria such as B. stearothermophilus may have grown and died within 5–7 days. Frequent examination of the incubated samples is therefore essential and it may also be prudent to examine destructively a proportion of the samples at intervals during longer incubation periods.

2.9.2 Mesophilic bacteria

Spore-forming bacteria These bacteria when present alone are usually survivors of the heat process, machine or equipment sterilisation (e.g. UHT sterilisers) or packaging decontamination (aseptic filling operations). Incubation at temperatures of 30–37°C are recommended for their growth, the time of incubation being longer at the lower temperature.

Incubation regimes of 14 days at 37°C, or 21 days at 30°C are frequently used.

Non-spore-forming bacteria These bacteria are usually associated with recontamination of surfaces, packaging or product and are often water or air borne. Many species are not able to survive, let alone grow, at temperatures in excess of 30°C and therefore incubation is recommended to be at temperatures of 20–30°C. Incubation times should be related to temperature, the lower the temperature the longer the incubation. It is therefore often not desirable to use the lower end of the incubation temperature range.

Incubation regimes of 14 days at 30°C, or 21 days at 25°C are frequently used.

Care must be taken if incubating at 'room temperature' since temperature cycling, especially in winter and at weekends, may be excessive. In extreme cases, organisms whose growth is initiated during warmer periods may be killed off during cold periods.

2.9.3 *Yeasts and microfungi*

These organisms may be survivors of pasteurisation treatment, in the cases of non-low-acid foods, or may be associated with recontamination of surfaces, packaging or product. Temperature requirements for growth are frequently below 30°C and the most commonly recommended incubation temperature is 25°C.

The incubation time is very dependent upon the nature of the product, acidity and water activity, and the type of organism which might be present. Some microfungi particularly, surviving heat pasteurisation in acid products, may require incubation for a minimum of 30 days to encourage sufficient growth to establish their presence.

It is recommended that a minimum period of 14 days at 25°C be used for all products and that longer periods of incubation time may be necessary for acid products as experience dictates.

References

1. Hersom, A.C. and Hulland, E.D., *Canned Foods. Thermal Processing and Microbiology*, 7th edn., Churchill Livingstone, Edinburgh, 1980.
2. Adam, W.B. and Dickinson, D., Scientific Bulletin No. 3, Campden Food and Drink Research Association, Chipping Campden, Glos, UK, 1959.
3. Pflug, I.J. and Holcomb, R.G., *Disinfection, Sterilisation and Preservation*, ed. S.S. Block, Lea and Febiger, Philadelphia, PA, 1977.

4. Brown, K.L. and Ayres, C.A., *Developments in Food Microbiology*, 1st edn., ed. R. Davies, Elsevier Applied Science, London, 1982, Chap. 4.
5. Stumbo, C.R., *Thermobacteriology in Food Processing*, Academic Press, New York, 1973.
6. Bigelow, W.D. and Esty, J.R., *J. Infect. Dis.* 27 (1920) 602.
7. Schmidt, C.F., *J. Bacteriol.* 59 (1950) 433.
8. Esty, J.R. and Williams, C.C., *J. Infect. Dis.* 34 (1924) 516.
9. Franklin, J.G., Williams, D.J. and Clegg, L.F.L., *J. Appl. Bacteriol.* 21 (1958) 51.
10. Resende, R., Stumbo, C.R. and Francis, F.J., *Food Technol.* 23 (1969) 325.
11. Perkin, A.G., Burton, H., and Davies, F.L., *J. Food Technol.* 12 (1977) 131.
12. Davies, F.L., Underwood, H.M., Perkins, A.G. and Burton, H., *J. Food Technol.* 12 (1977) 115.
13. Neaves, P. and Jarvis, B., Research Report No. 280, BFMIRA, Leatherhead, Surrey, England, 1978.
14. Neaves, P. and Jarvis, B., Research Report No. 286, BFMIRA, Leatherhead, Surrey, England, 1978.
15. Reichart, O., *Acta Aliment.* 7 (1978) 396.
16. Reichart, O., *Acta Aliment.* 8 (1979) 131.
17. Michiels, L., *Ind. Aliment. Agric.* 89 (1972) 1349.
18. Stern, J. A. and Proctor, B. E., *Food Technol.* 8 (1954) 139.
19. Cerf, O., Grosclaude, G. and Vermiere, D., *Appl. Microbiol.* 19 (1970) 696.
20. Brown, K. L., Aseptic packaging of vegetable products, Reports, Campden Food Preservation Research Association, Chipping Campden, Glos, UK, 1974, 1975.
21. Thorpe, R.H. and Grey, K.A., Technical Note 138, Campden Food Preservation Research Association, Chipping Campden, Glos, UK, 1971.
22. Perkin, A.G., Davies, F.L., Neaves, P., Jarvis, B., Ayres, C.A., Brown, K.L., Falloon, W.C., Dallyn, H. and Bean, P., *Microbial Growth and Survival in Extremes of Environment*, eds. G.W. Gould and J.E.L. Corry, Academic Press, London, 1980.
23. Pflug, I.J. Smith, G., Holcomb, R. and Blanchett, R., *J. Food Prot.* 43 (1980) 119.
24. Townsend, C.T., Esty, J.R. and Baselt, F.C., *Food Res.* 3 (1938) 323.
25. Sognefest, P., Hays, G.L., Wheaton, E. and Benjamin, H.A., *Food Res.* 13 (1948) 400.
26. Reynolds, H., Kaplan, A.M., Spencer, F.B. and Lichtenstein, H., *Food Res.* 17 (1952) 153.
27. Williams, C.C., Merrill, C.M. and Cameron, E.J., *Food Res.* 2 (1937) 369.
28. Cuncliffe, H.R., Blackwell, J.H., Dors, R. and Walker, J.S., *J. Food Prot.* 42 (1979) 135.
29. Galesloot, Th.E., *Neth. Milk Dairy J.* 10 (1956) 79.
30. Williams, D.J., Franklin, J.G., Chapman, H.R. and Clegg, L.F.L., *J. Appl. Bacteriol.* 20 (1957) 43.
31. Aiba, S. and Toda, K., *Proc. Biochem.* 2 (1967) 35.
32. Kooiman, W.J., *Spore Research 1973*, eds. A.N. Barker, G.W. Gould and J. Wolf, Academic Press, London, 1974.
33. Wang, D.I.C., Scharer, J. and Humprey, A.E., *Appl. Microbiol.* 12 (1964) 451.
34. Oquendo, R., Valdivieso, L., Stahl, R. and Loncin, M., *Lebensm.-Wiss. Techol.* 8 (1975) 181.
35. Stumbo, C.R., *Food Technol.* 2 (1948) 228.
36. Pflug, I.J. and Esselen, W.B., *Food Technol.* 7 (1953) 237.
37. Pflug, I.J., *Food Technol.* 14 (1960) 483.
38. Pflug, I. J. and Esselen, W.B., *Food Res.* 19 (1954) 92.
39. Secrist, J.L. and Stumbo, C.R., *Food Res.* 23 (1958) 51.
40. Frank, H.A. and Campbell, L.L. Jr., *Appl. Microbiol.* 5 (1957) 243.
41. Youland, G.C. and Stumbo, C.R., *Food Technol.* 7 (1953) 286.
42. David, J.R.D. and Shoemaker, C.F., *J. Food Sci.* 50 (1985) 674.
43. Brown, K.L., Gaze, J.E., McClement, R.H. and Withers, P., *Int. J. Food Sci. Technol.* 23 (1988) 361.
44. Perkin, A.G., *J. Dairy Res.* 41 (1974) 55.
45. Burton, H. and Perkin, A.G., *J. Dairy Res.* 37 (1970) 209.
46. Edwards, J.L. Jr., Busta, F.F. and Speck, M.L., *Appl. Microbiol.* 13 (1965) 851.
47. Busta, F.F., *Appl. Microbiol.* 15 (1967) 640.
48. Lindgren, B. and Swartling, P., Milk and Dairy Res. Rep. No. 69, Alnarp, Sweden, 1963.
49. Jacobs, R.A., Kempe, L.L. and Milone, N.A., *J. Food Sci.* 38 (1973) 168.

50. Hunter, G.M., *Food Technol. Australia* 24 (1972) 158.
51. Dallyn, H., Falloon, W.C. and Bean, P.G., *Lab. Practice* 26 (1977) 773.
52. Brown, K.L., Ayres, C.A., Gaze, J.E. and Newman, M.E., *Food Microbiol.* 1 (1984) 187.
53. Mechura, F.J., *Proceedings IUFoST Symposium on Aseptic Processing and Packaging of Foods*, Tylosand, Sweden, Lund University and SIK, 1985.
54. Hersom, A.C. and Shore, D.T., *Food Technol.* 35 (1981) 53.
55. Grieme, L.E. and Barbano, D.M., *J. Food Prot.* 46 (1983) 797.
56. Burton, H., *Proc. Int. Conf. on UHT Processing and Aseptic Packaging of Milk and Milk Products*, Dept. Fd. Sci., N. Carolina State University, Raleigh, NC, 1979.
57. Heppell, N.J., *J. Food. Eng.* 4 (1985) 71.
58. Pflug, I. J., *Textbook for an Introductory Course in the Microbiology and Engineering of Sterilisation Processes*, Environmental Sterilisation Laboratory, Union Street, Minneapolis, 1987.
59. Schmidt, C.F., Thermal resistance of microorganisms, *Aseptics, Disinfectants, Fungicides, and Chemical and Physical Sterilisation*, ed. G.F. Reddish, Lea and Febiger, Philadelphia, PA, 1954, pp. 831–883.
60. Ball, C.O., *Bull. Natl. Res. Council* 7 (1923) part 1 no. 37.
61. Pflug, I.J. and Christensen, R., *J. Food Sci.* 45 (1980) 35.
62. Mansfield, T., *Proc. 1st International Congress in Food Science and Technology*, Vol 4, Gordon and Breach, London, 1962, p. 311.
63. Holdsworth, S.D., *J. Food Eng.* 4 (1985) 89.
64. Shapton, D.A., *Process Biochem.* 1 (1966) 121.
65. Shapton, D.A., Lovelock, D.W. and Laurita-Longo, R., *J. Appl. Bacteriol.* 34 (1971) 491.
66. DHSS, Food Hygiene Code of Practice No. 10, *The Canning of Low Acid Foods. A Guide to Good Manufacturing Practice*, HMSO, 1981.
67. Food Processors Institute, *Canned Foods Principles of Thermal Process Control, Acidification and Container Closure Evaluation*, Food Processors Institute, Washington, 1980.
68. Campden Food Preservation Research Association, Guidelines for the establishment of scheduled heat processes for low acid foods, CFDRA Tech. Manual No. 3, Campden Food Preservation Research Association, Chipping Campden, Glos, UK, 1977.
69. Atherton, D. and Thorpe, R.H., The processing of canned fruit and vegetables, Tech. Bull. 4, revised edition, Campden Food Preservation Research Association, Chipping Campden, Glos, UK, 1980.
70. Codex Alimetarius, Vol. G, *Recommended International Code of Practice for Low-acid and Acidified Low-acid Canned Foods*, CAC/RCP23-1978 F.A.O. World Health Organisation, Rome, 1983.
71. National Canners Association, *Laboratory Manual for Food Canners and Processors*, Vol. 1—Microbiology and Processing, AVI, Westport, CT, 1968.
72. National Canners Association, Processes for low-acid canned foods in metal containers, Bulletin 26-L, NCA, Washington, 1976.
73. National Canners Association, Processes for low-acid canned foods in glass containers, Bulletin 30-L, NCA, Washington, 1971.
74. Lopez, A., *A Complete Course in Canning Book II: Processing Procedures for Canned Food Products*, Canning Trade Inc., Baltimore, MD, 1981.
75. Food and Drug Administration, 21 CFR Part 113, *Thermally Processed Low-acid Foods Packaged in Hermetically Sealed Containers*, F.D.A., Washington, 1979.
76. Food and Drug Administration, 21 CFR Part 114, *Acidified Foods*, F.D.A., Washington, 1979.
77. The Milk (Special Designation) Regulations 1977, Statutory Instrument No. 1033, HMSO, London, 1977.
78. The Milk and Dairies (Heat Treatment of Cream) Regulations 1983, Statutory Instrument No. 1509, HMSO, London, 1983.
79. The Milk-Based Drinks (Hygiene and Heat Treatment) Regulations 1983, Statutory Instrument No. 1508, HMSO, London, 1983.
80. Recommended Guidelines for Controlling Environmental Contamination in Dairy Plants, revised Oct 1987, issued jointly by the U.S. Food and Drug Administration and Milk

Industry Foundation International Ice Cream Association, *Dairy and Food Sanitation* 8 (2) (1988) 52–56.
81. *Guidelines for Good Hygienic Practice in the Manufacture of Soft and Fresh Cheeses*, Creamery Proprietors Association, 19 Cornwall Terrace, London, NW1 4QP, 1988.
82. Slocum, G.G., Welsh, H. and Hunter, A.C., *Food Res.* 6 (1941) 179.
83. Stumbo, C.R., Murphy, J.R. and Cochran, J., *Food Technol.* 4 (1950) 321.
84. Gaze, J.E. and Brown, K.L., *Int. J. Food Sci. Technol.* 23 (1988) 373.
85. Scott, V.N. and Bernard, D.T., *J. Food Prot.* 45 (1982) 909.
86. Ito, K.A., Seslar, D.J. and Mercer, W.A., *Botulism 1966*, eds. M. Ingram and T.A. Roberts, Chapman and Hall, London, 1967, pp. 108–122.
87. Lynt, R.K., Kautter, D.A. and Solomon, H.M., *J. Food Sci.* 47 (1981) 204.
88. Lynt, R.K., Kautter, D.A. and Solomon, H.M., *J. Food Sci.* 44 (1979) 108.
89. Lynt, R.K., Solomon, H.M. and Kautter, D.A., *J. Food Prot.* 47 (1984) 463.
90. Pflug, I.J. and Odlaug, T.E., *Food Technol.* 32 (1978) 63.
91. Esty, J.R. and Meyer, K.F., *J. Infect. Dis.* 31 (1922) 650.
92. Center for Disease Control, Botulinum in the United States 1899–1973, *Handbook for Epidemiologists, Clinicians and Laboratory Workers*, DHEW Publ. (CDC) 74-8279, US Dept. of Health, Education and Welfare, Atlanta, GA, 1974.
93. Anon, *Food Chem. News* Oct 31 (1983) 19.
94. Varela, G., Bender, A.E. and Morton, I.D., eds., *Frying of Food, Principles, Changes, New Approaches*, Ellis Horwood, Chichester, UK, 1988.
95. Brown, K.L., *J. Appl. Bacteriol.* 65 (1988) 49.
96. Russell, A.D., *The Destruction of Bacterial Spores*, Academic Press, London, 1982.
97. Esselen, I.J. and Pflug, I.J., *Food Technol.* 10 (1956) 557.
98. Secrist, J.L. and Stumbo, C.R., *Food Technol.* 10 (1956) 543.
99. Matsuda, N., Komaki, M., Ichikawa, R. and Gotoh, S., *Nippon Shokuhin Kogyo Gakkaishi* 32 (1985) 391.
100. Stringer, M.F., *Clostridium in Gastrointestinal Disease*, ed. S. Borriello, CRC Press, Florida, 1985.
101. Hobbs, B.C., *Handbook of Foodborne Diseases of Biological Origin*, ed. M. Rechcigl, Jr., CRC Press, Florida, 1983.
102. Gaze, J.E., Brown, K.L., Brown, G. and Stringer, M.F., Technical Memorandum 466, Campden Food and Drink Research Association, Chipping Campden, Glos, UK, 1987.
103. Knock, G.G., Lambrechts, M.S.J., Hunter, R.C. and Riley, F.R., *J. Sci. Food. Agric.* 10 (1959) 337.
104. Parry, J.M. and Gilbert, R.J., *J. Hyg. Camb.* 84 (1980) 77.
105. Donnelly, L.S. and Busta, F.F., *Appl. Environ. Microbiol.* 40 (1980) 721.
106. Brown, K.L., Fundamental and applied aspects of spores, *Proceedings of Cambridge Spore Conference*, Academic Press, London, 1983, pp. 387–394.
107. Xezones, H., Segmiller, J.L. and Hutchings, I.J., *Food Technol.* 19 (1965) 1001.
108. Segner, W.P., *Food Technol.* 33 (1979) 55.
109. Matsuda, N., Masuda, H., Komaki, M. and Matsumoto, N., *J. Food Hyg. Soc. Japan* 23 (1982) 480.
110. Doores, S. and Westhoff, D., *J. Food Sci.* 46 (1981) 810.
111. Tomlins, R.I. and Ordal, Z.J., *Inhibition and Inactivation of Vegetative Microbes*, eds. F.A. Skinner and W.B. Hugo, SAB series No. 5, Academic Press, London, 1976.
112. Mackey, B.M. and Bratchell, N., *Lett. Appl. Microbiol.* 9 (1989) 89.
113. Gaze, J.E., Brown, G.D. and Banks, J.G., *Food Microbiol.* (1989) in press.
114. Magnus, C.A., McCurdy, A.R. and Ingeledew, W.M., *Can. Inst. Food Sci. Technol. J.* 2 (1988) 209.
115. Put, H.M.C. and de Jong, J., *J. Appl. Bacteriol.* 52 (1982) 235.
116. Taylor, D.M., *PHLS Microbiol. Digest* 5 (1) (1988) 14.
117. Guzewich, J.J. and Morse, D.L., *J. Food Prot.* 49 (1986) 389.
118. West, P.A., Wood, P.C. and Jacob, M., *J. R. Soc. Health* 105 (1985) 15.
119. Fassolitis, A.C., Peeler, J.T., Jones, V.I. and Larkin, E.P., *J. Food Prot.* 48 (1985) 4.
120. Filppi, J.A. and Banwart, G.J., *J. Food Sci.* 39 (1974) 865.

Further reading

Anon, *Dairy Food* Sanitation 6(4) (1986) 166.

DHSS, Management of patients with spongiform encephalopathy (Creutzfeldt-Jacob disease CJD), DHSS circular DA (84) 16, 1984.

Halvorson, H.O. and Zeigler, N.R., *J. Bacteriol.* 25 (1933) 101.

Larkin, E.P., *CRC Handbook of Foodborne Diseases of Biological Origin*, ed. M. Rechcigl, Jr., CRC Press, Florida, 1983, Chap. 1.

The IDF definitions of heat treatment as applied to milk and milk products, D-DOC 170, 1988.

3 Heat processing equipment

P.S. RICHARDSON and J.D. SELMAN

3.1 Introduction

The thermal processing operation requires the heating of food products through a predetermined time and temperature cycle in order to render the contents commercially sterile; that is free from micro-organisms likely to cause food poisoning and with the level of spoilage organisms reduced to a commercially acceptable level. For a low acid food product, one of pH > 4.5, this process requires the heating of the product to temperatures above 100°C usually in the range 115–130°C for a time sufficient to achieve a 12 log reduction of the spores of *Clostridium botulinum* as defined in Department of Health Code of Practice No. 10 [1]. Current practices are, however, to move to even higher temperatures and consequently shorter process times so maximising the organoleptic and nutrient retention within the product. The time–temperature procedure required to render a product commercially sterile must be carefully determined using established procedures [2]. Thermal processes may be achieved after the product has been filled and sealed in containers, e.g. by canning or in continuous flow systems prior to aseptic packaging (e.g. UHT processes). In this chapter the operating principles of a range of thermal processing systems are discussed together with a consideration of the newer techniques which are still at various stages of development.

3.2 In-container processing

The food processing industry produces a wide range of products in a variety of containers necessitating the need for an equally wide range of processing techniques and hence retort designs and operating procedures.

Retorting systems can be subdivided in several ways. The main distinction comes between continuous retorting systems, that is those in which containers are continuously fed into and out of the retort, and batch systems in which the retort is filled with product, closed and then put through a processing cycle. Further classification can be made on the type of heating medium that is used in the retort; steam, steam/air or water. Batch retorts are available in a number

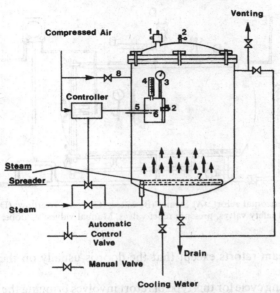

Figure 3.1 A vertical batch retort. (1) Safety valve; (2) petcocks to maintain a steam bleed from retort during processing; (3) pressure gauge; (4) thermometer; (5) sensing element for controller; (6) thermo-box; (7) steam spreader; (8) air inlet for pressure cooling.

of configurations for various applications including static, rotary, steam heated and water heated with or without air overpressure. The air overpressure is necessary to maintain the integrity of the containers during retort operating cycles for glass and flexible containers.

3.2.1 Batch systems

Batch steam retorts These are usually arranged either vertically (Figure 3.1) or horizontally (Figure 3.2) and are used for canned products which are placed into baskets immediately after seaming, and are then placed into the body of the retort shell. The retort comprises a metal shell pressure vessel which is fitted with inlets for steam, water and air and has outlet ports for venting air during retort bring up, and for draining the retort at the end of the cycle. A pocket for instruments, thermometer, temperature recording probe and pressure gauge, is located on the side of the vessel. To ensure adequate steam movement around the instruments, in particular the temperature sensors, the instrument pocket is fitted with a constant steam bleed. On vertical retorts the lid is hinged at the top and secured to the shell during processing by several bolts. One bolt is usually a safety bolt designed to allow the venting of any excess pressure which may have been inadvertently left in the retort at the end of the process when the lid is released. A similar arrangement is used on

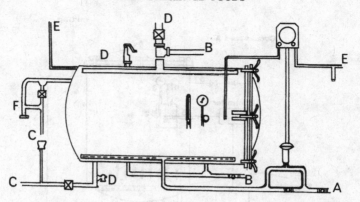

Figure 3.2   A horizontal retort. (A) Steam; (B) water; (C) drain, overflow; (D) vents, bleeder; (E) air; (F) safety valves, pressure relief valves. Manual valves: o, globe; ⊠, gate.

horizontal steam retorts except that the door is usually on the end of these machines.

The operating cycle for this type of retort involves bringing the retort up to a temperature of around 100°C, and then allowing steam to pass through the vessel to the atmosphere for sufficient time so that all air in the retort and between the cans is removed (venting) before the retort is finally brought up to the operating pressure and processing temperature. At the end of the processing time, the steam is turned off and a mixture of cooling water and air introduced into the retort to cool the cans. The purpose of the air is to maintain the pressure in the retort following the condensation of the residual steam after the initial introduction of cooling water. If this pressure is not maintained the containers may deform due to pressure imbalances between the internal pressure in the cans and the retort. As the temperature drops, the pressure in the retort may be controlled and gradually reduced until atmospheric pressure is reached and water can be allowed to flow through the retort, cooling the cans to a temperature of about 40°C before they are removed from the retort. Cans are removed from the retort at this temperature since this allows the surface of the cans to dry rapidly by evaporation so reducing the risk of leaker spoilage [3]. The water is preferably sprayed or alternatively the retort may simply be filled and allowed to stand for sufficient time for the cans to cool to the unloading temperature.

Both of these systems are static in operation. For other types of product it is possible to assist the rate of heat penetration by agitating the cans in the steam environment by rotation either about the horizontal axis in a horizontal retort or by rotation in the vertical plane in a vertical retort. This has the benefit of reducing the over processing that would occur for conduction heating products at the surface of the product in order to achieve the desired thermal process at the slowest heating point.

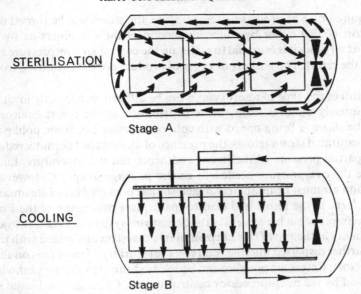

Figure 3.3 The Lagarde system. Stage A: the sterilising process uses forced steam circulation. Stage B: the rapid cooling is achieved by spraying of recycled cold water.

Some steam retorts incorporate a fan to force the steam through the retort to increase the rate of heat transfer from steam to the containers. An example of this type of system is the Lagarde system illustrated in Figure 3.3.

Steam/air retort systems   The use of glass and plastic containers has increased the use of alternative retorting systems. With these types of containers it is usually not sufficient to rely on the strength of the containers alone to counteract the build up of internal pressure during heating, but a constant overpressure of air is required to ensure the integrity of the package during heating. Thus the heating medium used in this type of retort is often a mixture of steam and air in proportions designed to provide the necessary steam temperature and air overpressure to maintain the package integrity. In order to ensure adequate mixing of the steam and air these retorts are fitted with a fan system to disperse the steam and air, so eliminating the possibility of the development of cold spots in the processing chamber.

The containers are loaded into crates or cages taking care to ensure good contact between the heating medium and the products to be sterilised. These systems are usually batch in operation, requiring the filled and sealed containers to be placed in the processing cages which are then placed into the retort from one end through a door. The retort will then be heated to processing temperature with sufficient air overpressure being applied to maintain the integrity of the packs in the retort undergoing processing. After the required time at processing temperature, the steam will be turned off and the retort cooled either by water immersion of the containers or by water spray. At all times it is essential to maintain the correct air overpressure regime within the retort. The baskets or cages of product are then removed from the retort.

Control of this type of retort system can be difficult, particularly in ensuring an adequately uniform temperature distribution in the retort environment when the steam is being mixed with cold compressed air. Here, unlike in the case of saturated steam retorts, the presence of air must not permit a reduction in the partial pressure of the steam and hence retort temperature, but only provide the overpressure needed to ensure package integrity. However, the steam and air must be intimately mixed so that pockets of cold steam/air mix do not form in the retort and lead to inadequate processing of the cans.

An example of such a system is the Steristeam system. Steam is introduced to the cavity at the top of this horizontal retort system and mixed with the air, being further dispersed into the retort by the centrifugal fan or fans on the side of the retort. Excess air pressure can be released through the air vent which is controlled by the microprocessor control system. Coolings is achieved either by injection of cooling water from the treated reservoir or by the use of chilled air. Cooling water can be recycled.

Water processing retorts This style of retort system has shown a rapid growth in recent years. The first major class of retort to use this heating medium was that used for the processing of glass jars. They were static batch retorts into which baskets of glass jars were placed, immersed in warm water and then heated to processing temperature by the injection of live steam into the retort. The water was recirculated in order to prevent the development of hot and cold regions. The water in this type of retort should be recirculated several times per minute to ensure a uniform temperature distribution. Current practices with semi-rigid containers utilise both water immersion and raining water techniques with or without rotation. For the processing of semi-rigid containers, it is necessary to consider the arrangement of the packs in the appropriate retort baskets or cages so that sufficient water flow occurs to heat the products to sterilisation temperatures during the thermal processing cycle.

Raining water techniques (Figure 3.4) require the use of either an external steam injection system or heat exchanger system outside the direct environment of the retort. In the latter case the cold water feeding the system is combined with the recycled heating medium and raised to the temperature required in the retort before being admitted to the sterilisation chamber through a spray arrangement. The containers will have been arranged to allow good contact between the hot water heating medium and the product either using spacer bars or distribution plates. It is imperative that a good distribution of the water occurs as otherwise stratification may occur and certain products will receive an inadequate heat process. Control of the

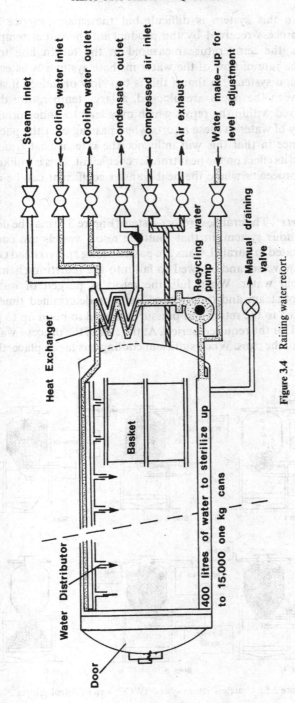

Figure 3.4 Raining water retort.

temperature in this system is difficult but the safest practice is to base the thermal process received by the product on the outlet temperature of the retort, i.e. the temperature measured in the return line to the heat exchanger. The rate of flow of the water in these systems is as critical as in water immersion systems, in that if this is too slow or indeed if some of the channels between the packs are blocked, a large temperature distribution may be observed within the retort which may lead to under processing.

The velocity of water in these retorts when passing over the packages is of vital importance in that this will influence the rate of heat transfer to the product due to its effect on the heat transfer coefficient. This is unlike saturated steam retort processes where the heat transfer coefficient can be considered infinite [4].

Crateless retorts   The crateless retort system (Figure 3.5) may be described as a semi-continuous system, in that multiple retort vessels are continuously filled and operated in parallel. Cans are passed along an overhead conveyor to the appropriate vessel and allowed to fall into the retorting chamber into a cushion of cold water. When full, the retort is purged of water by the introduction of steam and then vented for a predetermined time. Steam is further admitted to the retort and pressure allowed to build up to processing levels and held for the required period. At the end of the process, water and air are used to cool the cans. When sufficient cooling has taken place, the retort is

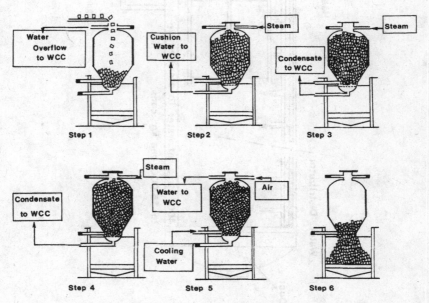

Figure 3.5 Crateless retort system (WCC; water control centre).

allowed to fill with water and the bottom door opened to allow the cans to enter the cooling canal. Systems usually have five or six processing vessels, often termed pots. One of the major advantages of this system is that it can handle large volumes of product similar to fully continuous systems but also allows the flexibility in terms of process selection that is achievable with batch systems.

3.2.2 *Continuous retort systems*

For large throughputs of canned products, batch retorting is not cost effective so continuous retort systems have been developed. Two major types are commonly used, namely the hydrostatic retort and the reel and spiral or cooker/cooler system. Other systems exist such as the Hydrolock retort, also semi-continuous systems such as the Universal.

Hydrostatic retort systems This type of system is illustrated in Figure 3.6. Cans are formed into 'sticks' and continuously fed into the retort onto carrier bars. The carrier bars then rise and fall through the water legs to regions of higher pressure before entering the steam chamber where the cans are heated to and held at processing temperature for sufficient time to render the product commercially sterile before exiting the steam chest into the cooling legs of the

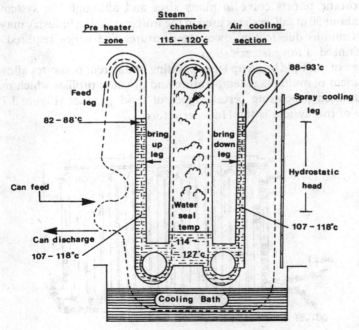

Figure 3.6 Hydrostatic retort system.

retort systems. Again use is made of hydrostatic legs to gradually reduce the pressure on the outside of the cans to atmospheric pressure. The cans are initially spray cooled, then immersion cooled before exiting the retort system for further labelling and packing. Within a hydrostatic retort system there is a limited degree of agitation to assist the heat transfer since the cans change direction as they go over the top of the columns and similarly at the base of the columns.

The position of the water seal in the steam chest is critical and is carefully controlled as this marks the point where cans enter and leave the steam heating region of the retort, i.e. where the sterilisation process takes place. Like all retort systems the measurement of time and temperature is central to the determination of the thermal process received by the cans and to this end the temperature in the steam chest must be carefully monitored. It is usual to have several thermometers mounted at different levels in the steam chamber and in addition monitor the temperature in all other legs of the retort system. Process time is monitored by controlling the speed of the drive chain on which the carrier bars are mounted. In order to achieve different process values, the speed of the drive chain on these retorts can be varied so controlling the length of time that the cans spend in the steam processing chest [5].

These retort systems may also include multiple feed points to take cans from different production lines at the same time. Similarly multiple discharge points may be used.

Hydrostatic retorts come in many sizes and although the systems are usually about 30 m tall, systems used in the fruit processing industry may have shorter columns due to the lower temperatures and hence required over-pressures needed for pasteurisation.

The use of multiple bring up legs operating at different pressures allows the development of multi-stage temperature and pressure profiles which may be required when processing certain types of food product (Figure 3.7). An example of this system is the Hunister retort.

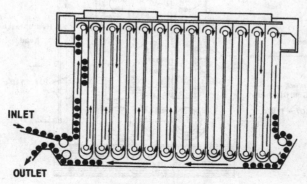

INLET

OUTLET

Figure 3.7 The Hunister retort.

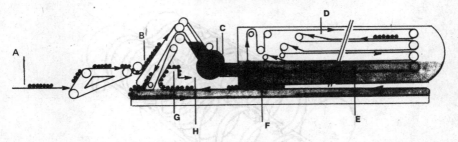

Figure 3.8 The Hydrolock retort system. (A)–(B) Loading; (C) waterlock valve; (D) steam sterilisation area; (E) pressurised water, cooling area; (F) atmospheric water, cooling area; (G)–(H) unloading. Shaded area marks the cooling section.

The Hydrolock system   The Hydrolock retort system allows the continuous processing of cans in a steam/air environment in a similar way to the steam heated hydrostatic retort, except that the cans are introduced into, and removed from the pressurised steam environment through a pressure lock (Figure 3.8). Cans are placed into a carrier chain before passing through the pressure lock. The chain passes through a water seal into the steam environment where it makes multiple passes in the fan circulated steam/air chamber in order to ensure an adequate thermal process. After this the chain is submerged under the water seal to allow precooling before being passed back through the pressure lock to be air cooled at atmospheric pressure prior to discharge from the retort system.

Reel and spiral retort systems   This second type of continuous retort system differs from the hydrostatic system in that it allows the continuous processing of cans with some rotation which agitates the can contents so bringing about the benefits of more rapid heat transfer. The system is illustrated in Figure 3.9. Each chamber is linked to the other by a transfer valve mechanism which may include a pressure lock. The first shell of the system is at atmospheric pressure and used for preheating the cans. The next shell is at higher pressure and it is here that the thermal process is delivered before the cans pass to the cooling shells. These reduce the can temperature but allow the maintenance of air overpressure as appropriate to maintain the can shape during the cooling operation. The heating medium is usually steam but water immersion systems have been developed [6]. As the cans enter the system, they are fed into a position on a rotating reel which rotates at up to 6 rev./min. As the reel rotates, the cans are driven through the retort system by a spiral attached to the inside of the shell so allowing a helical motion with rotation through the retort shell. To ensure synchronisation of the shell drives, all are driven by one motor and linked by intermeshing gears. The system is easily upset by damaged

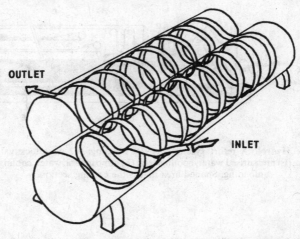

Figure 3.9   Path of cans through a reel and spiral cooker/cooler.

containers, so such systems are usually preceded by can inspection equipment. In the event of a damaged can causing a jam in the system, a shear pin is incorporated in the drive system to prevent further damage occurring to the motor system. The cans do not rotate at a constant rate around all parts of the shell. As can be seen in Figure 3.10, the cans are carried by the reel for about 60% of the rotation and only rotate within the reel as they traverse the shell at the bottom. Therefore the maximum benefit of the rotary processing technique is only achieved in this 30% of the retort.

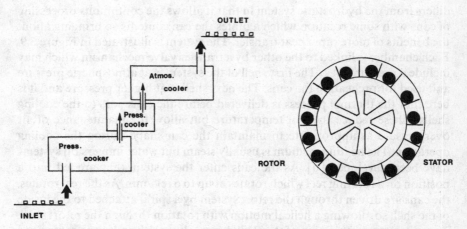

Figure 3.10   (a) Three shell line arrangement; (b) axial rotary movement of cans in a continuous steriliser.

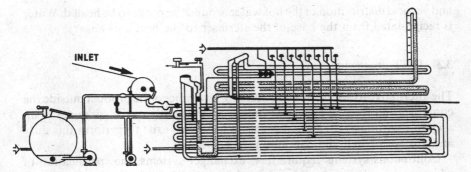

Figure 3.11 The hydroflow system.

The hydroflow system This retort system is illustrated in Figure 3.11. The major difference in this system from other continuous retort systems is that the cans are conveyed through the equipment by water, with rotation of the cans about their horizontal axis being induced by the cans rolling along the guide rail. The cans pass through preheat, sterilisation, precool and cooling sections within the retort.

The universal steriliser This retort system (Figure 3.12) allows the automation of the processing of containers, glass or plastic, packed in baskets or cages. The filled baskets or cages are continuously trucked to the front of this horizontal system which is equipped with a pressure lock at each end. The heating medium is hot water which is heated externally in a heat exchanger. Each truck is admitted into the retort through the twin gated pressure lock and then conveyed by a positive drive screw through the heating and pressure cooling phases of the operation, before emerging through the twin gated pressure lock at the opposite end of the retort for final cooling and draining of the cans before unloading. Hot water is incident on the containers from the top but adequate distribution plates, used as layer pads, ensure both horizontal

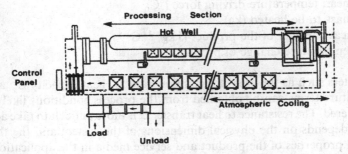

Figure 3.12 The universal steriliser.

and vertical distribution of the hot water around the packs to be heated. Water is recirculated from the base of the steriliser to the heat exchanger.

3.3   UHT processing

The alternative to in-container processing is to process the product outside the container and then fill and seal it into a commercially sterile container in an aseptic environment. Thermal processing systems exist to perform this duty and are both batch and continuous in operation.

Continuous systems require heat exchanger systems, the specification of which is chosen after consideration of the type of product to be heated and its likely rheological and physical properties during processing. The heat exchanger system is only used to raise the product to and cool it from the processing temperature. The lethal effect of the thermal process occurs and is controlled in the holding section of the continuous flow plant or in the holding period of a batch operation. The specification and operation of such systems has been addressed by Rose [7]; see also Chapter 4.

3.3.1   *Heat exchangers*

The heat exchanger is designed to allow a product to be heated or cooled using another service medium as the heat source or sink. The most common heat transfer medium for heating applications is steam, and for cooling applications, water. In essence two expressions can be used to describe the heat transfer operation:

$$Q = U \times A \times T_m = M \times C_p \times T$$

where
  Q = rate of heat transfer (J s^{-1})
  U = heat transfer coefficient, a measure of the resistance to heat transfer
       (J m^{-2} S^{-1} °C^{-1})
  A = area available for heat transfer (m^2)
  T_m = mean temperature driving force (°C)
  M = mass to be heated (kg)
  C_p = heat capacity of the product (J kg^{-1} °C^{-1})
  T = temperature change of product (°C)

Parameters such as the area A, mass flow rate M, heat capacity C_p, and the temperature terms can be obtained from the process conditions likely to be encountered. The resistance to heat transfer U is more difficult to take account of as it depends on the physical dimensions of the system and the thermo-physical properties of the product and service media in the application. The resistance to heat transfer may change with time in a given process, for example if a fouling layer of burnt-on product were allowed to accumulate on the heat transfer surface.

The resistance to heat transfer in conduction heating situations is easy to visualise, but in convection heating situations has classically been considered to manifest itself in a layer immediately adjacent to the heat transfer surface, with contributing resistances being additive, 'making the imaginary layer thicker'.

3.3.2 *Designs of heat exchangers*

Many designs of heat exchanger exist using different geometries and materials. Continuous heat exchangers can be subdivided into two broad categories: indirect systems and direct systems. Indirect systems use a heat transfer surface to keep the heating, or cooling, media separated from the product. Examples are tubular, scraped surface and plate heat exchangers. Direct systems allow the product and heating media to mix intimately, such as steam injection and steam infusion systems. Cooling of products heated in this way is either achieved by evaporative cooling or by using an indirect system.

Tubular heat exchangers These consist of two or three concentric tubes, or a single tube within an outer casing (Figure 3.13). The mechanical strength of these tubes allows them to operate at high temperatures and pressures. The turbulence required to minimise the film resistance to convective heat transfer is provided only by the velocity of the product. It follows that at normal operating conditions, the velocity required to generate turbulence in thick or viscous products would not be achieved so this design of heat exchanger is limited in its application to thin products such as milk or juices.

Some manufacturers of tubular heat exchangers use corrugated tubes for certain applications in order to improve the heat transfer efficiency due to an increased surface area and also by an improvement in the level of turbulence which is induced for a given product velocity.

Scraped surface heat exchangers Scraped surface heat exchangers are a double tube heat exchanger with the addition of a scraping mutator shaft

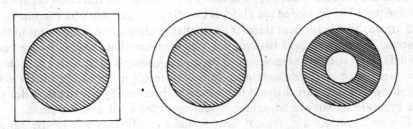

Figure 3.13 Different configurations for tubular heat exchangers.

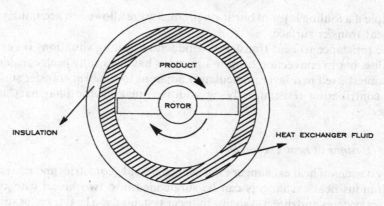

Figure 3.14 Scraped surface heat exchanger.

located within the product tube so that the rotating scraper blades keep the heat transfer surface clean (fouling may limit the resistance to heat transfer) and increase the mixing and turbulence within the heat transfer system (Figure 3.14). This type of heat exchanger is particularly useful for high viscosity products or products containing sizeable particulate material. Various arrangements of blades and diameters of mutator shaft are available so that a compromise between the requirements of product integrity and efficient heat transfer may be reached. The heating medium can be steam, or if a more gentle rate of heating is required (i.e. without the release of latent heat), hot water. Orientation of the heat exchangers may be either horizontal or vertical depending on the design.

Typical values for the overall heat transfer coefficient (U) are 1100–3400 $Wm^{-2}K^{-1}$, and shaft speed 100–500 rev./min.

Plate heat exchangers These consist of a series of plates, terminals between the plates and a head terminal onto which the plates are pressed with the end terminal. Product and heating, or cooling media, flow in alternate channels in thin layers (Figure 3.15). Sealing between the plates is usually by elastic sealing gaskets of synthetic rubber cemented into a preformed groove. The plates are of polished stainless steel approximately 0.5–1.25 mm in thickness separated by 3–6 mm. The surface of the plates is usually corrugated in order to increase the area available for heat transfer and enhance the turbulence present in the system. The narrow gaps between the plates means that these units are best suited to low viscosity homogeneous products. Attempts to process particulate products, such as juice containing cells, will result in blocked headers and resulting large pressure drops in the system and eventually blown plates due to the pressure imbalance between product and media sides of the plate.

For milk processing, typical heat transfer coefficients are in the range 1500 $Wm^{-2}K^{-1}$ for heating and 1000 $Wm^{-2}K^{-1}$ for cooling.

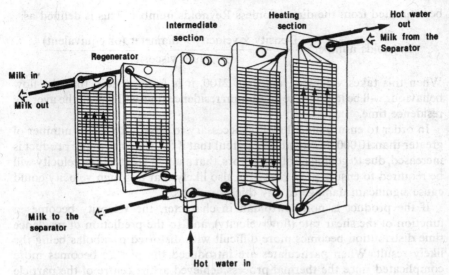

Figure 3.15 Product flow through a plate heat exchanger.

Direct heating systems These systems rely on intimate contact between the steam and the product. Heating is by condensation. The release of the large amounts of energy associated with the latent heat result in a rapid rise in temperature. Careful design of the injector systems can allow this technique to be applied to both high and low viscosity products but not to particulate products.

The condensing steam dilutes the product. As a rough guide, a watery product requires 0.2 lb kg/kg to heat it from 21°C to 132°C.

This water can be removed by evaporative, or flash, cooling. This requires that the hot product is passed through a throttling disc to an area of lower pressure where a proportion of the water in the product spontaneously boils. The latent heat required to vaporise the product comes from the product, and so results in a rapid cooling of the product.

3.3.3 Holding sections

The holding section is the part of a continuous flow sterilisation plant where the product is considered to receive the thermal process sufficient to render it commercially sterile. The product is held at constant temperature dictated by the controller at the entrance to the holding section, for a time dictated by the length of the holding section and the flow characteristics of the product. If the product is homogenous it is likely that it will behave according to normal fluid flow laws, that is it will give a parabolic velocity distribution for laminar flow and approach plug flow in turbulent conditions. The degree of turbulence can be estimated from the dimensionless Reynolds number. This is defined as:

$$\text{Reynolds number} = \frac{\text{Density} \times \text{velocity} \times \text{diameter (or equivalent)}}{\text{Fluid viscosity}}$$

When this takes values of less than 2100, it is fairly certain that the flow behaviour will be laminar, i.e. minimum residence time will be half the average residence time.

In order to ensure plug flow it is necessary to have a Reynolds number of greater than 10 000. It is readily apparent that if the viscosity of the product is increased, due to gelling or other factors, that a much greater flow velocity will be required to ensure plug flow. It is also likely that this high velocity could cause significant damage to any thick products.

If the product is non-newtonian in character, the viscosity becomes a function of the shear rate (flow velocity), and so the prediction of residence time distribution becomes more difficult with distorted parabolas being the likely result. When particulates are introduced the picture becomes more complicated since the thermal process achieved at the centre of the particle must be sufficient to render it commercially sterile [8,9].

Alternative designs are now appearing in relation to holding tubes for particulates, such as the Stork 'Rotorhold' which is a circular chamber in which there are spokes that rotate to retard the progress of particulates but allow the liquid to go through much faster. The effect is to reduce the degree of over processing received by the liquid but to ensure the integrity of the process received by the particulates.

3.3.4 Batch processing systems

APV Jupiter systems As an alternative to continuous flow systems, particularly for products which contain large particulates, a batch processing system for the UHT processing of large particulates has been designed by the A.P.V. Co. Ltd. and marketed as the 'APV Jupiter system'. This process permits the separate processing of liquid and solid phases to ensure that each phase receives the optimum heat treatment. The liquid phase would be processed in a conventional continuous heat exchange system: plate, scraped, or direct steam injection. The sterilised liquid phase would then be added to the sterile solid phase after processing. The central feature of the Jupiter system is the double cone aseptic processing vessel (DCAPV). The DCAPV is a rotating jacketed vessel with inlet and outlet trunnions through which the product and service fluids flow. Within the inlet trunnion, a stationary core contains an inlet pipe for the injection of steam, air and liquids, a movable drain and vent system and a thermometer dipole to continually monitor the solid bed and steam temperatures during processing. The draining system acts as a vent during sterilisation and can be lowered into the product to remove

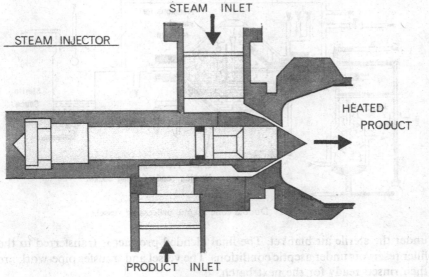

Figure 3.16 Steam injector.

any excess condensate and cooking liquid before the addition of the sterile sauce. The outlet trunnion houses the product pipe which runs from the outlet valve on the vessel via a rotary joint to the filler reservoir.

The Jupiter system is suitable for sterilising most vegetables and meat products. The addition of a cooking liquid reduces the risk of mechanical or thermal damage at temperatures above about 100°C when many products have softened and become fragile.

After loading, the solids are heated by direct steam injection in the vessel headspace and by indirect heating through the jacket while the vessel is rotating between 2 and 20 rev./min (Figure 3.16). The liquid feed and product pipe-work may be sterilised at this stage. Any gas evolved during this part of the process is vented via the drain and vent system.

During the heating cycle preheated liquid, water or stock, can be injected into the vessel to ensure optimum cooking conditions, and in some instances to protect fragile products from damage. Meanwhile sterilised product from the previous batch feeds from the reservoir to the aseptic fillers (Figure 3.17).

When the required heat treatment has been achieved, product is cooled by replacing steam with cooling water in the jacket and by admitting sterile air to the vessel and pipe-work. The sterile air is maintained at a small positive pressure during the remainder of the process cycle. At the end of the cooling cycle any excess liquid or condensate is removed from the stationary vessel via the drain system which has been lowered into the product bed. Separately sterilised sauce is added to the cooked product and mixed by vessel rotation

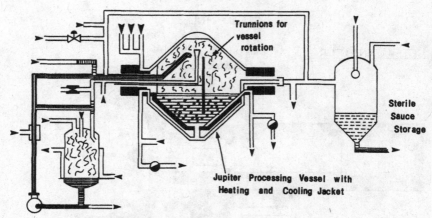

Figure 3.17 Double cone aseptic processing vessel.

under the sterile air blanket. The final blended product is transferred to the filler reservoir under aseptic conditions. The vessel and transfer pipe-work are then rinsed ready for the next batch.

Ohmic heating system   For UHT processing prior to aseptic filling, the most recent developments have included ohmic heating. This system requires that the food product acts as the dielectric in an electrical cell. With the ohmic heating system, heating rates can be greater than 50°C/s depending on the conductivity of the products.

The penetration depth with this type of heating is almost unlimited. Products are pumped through the system past a series of unevenly spaced electrodes (Figure 3.18) which contact the product allowing current to pass through it, so generating heat. The product then enters a holding section where the thermal process is delivered before being cooled by conventional indirect heat exchanger systems, the design of which is dictated by the product (usually tubular or scraped surface heat exchangers). Although the ohmic heating part of the system is capable of handling large particulate products with little damage, it must be remembered that the product must still be pumped and cooled by conventional means and it is these operations that may present the limiting factors to the applicability of the system together with the availability of suitable aseptic packaging systems.

3.4 Microwave heating systems

Microwave heating systems have been developed for both in-container processing and continuous flow processing systems. The majority of European systems utilise the 2450 MHz frequency and combine the effects of microwave heating and conventional heating techniques. Thus hot air or steam is coupled

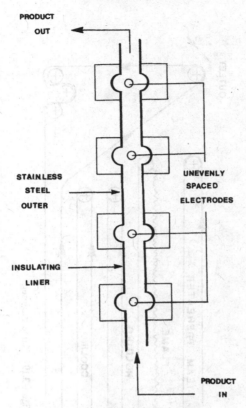

Figure 3.18 APV ohmic heater.

with microwaves to produce a rapid temperature rise and controlled temperature distribution at the surface and throughout the product. The in-container systems are usually in the form of a tunnel into which the filled and sealed containers are passed before being subjected to microwaves to bring them to the process temperature required. Most of these systems are aimed at producing pasteurised products which may require chilled distribution and storage [10]. Microwave in-container sterilisation processes are still being developed by several companies.

3.5 Flame sterilisation

Flame sterilisation is a method of high temperature short-time processing for filled, sealed cans. The cans pass over a series of gas burners, heating up very rapidly. This form of direct heating induces internal can pressures usually much higher than those associated with conventional in-container processing,

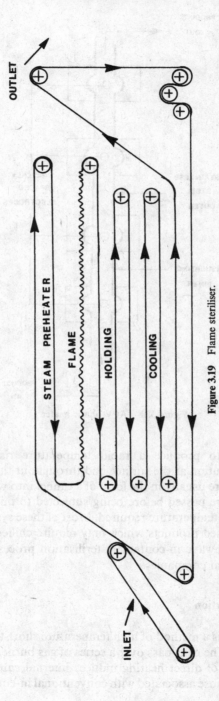

Figure 3.19   Flame steriliser.

with reference to process time and temperature of the product and the
temperature distribution at the critical point in the product. The hot
air/flame systems are used for sterilisation. Once it is heated and
sealed container are passed before cooling, allowing product moves to prior
than in the process the overall temperature heat. These systems are noted as
used continuous process which is widely scientific shed. In other cooling
sterilise 1101. Microswith no continuous sterilisation process, are still being
developed for several.

3.9   Flame sterilisation

Flame sterilisation is a means of heat storage which allows the process for
filled sealed cans. The principle of a series of cans heat by moving flame til sea
moult. This type of direct heating with very short time cup basically usually
much higher than those associated with conventional in-container processing

the cans not being subject to an external balancing counter pressure provided by the heating medium in a conventional retort.

Flame sterilisation has been limited to cans of less than 90 mm diameter. The ends of these cans often exhibit a modified profile and heavier gauge material to offset the greater pressures generated. The products processed by flame sterilisation must be 'mobile', i.e. convection heating packs. The cans are closed at high vacuum (> 50 cmHg) again to offset the pressures generated in the can.

The process has been used to advantage for the canning of mushrooms where less product shrinkage has been observed when compared with conventionally processed product. The system (Figure 3.19) is currently not in widespread use [11]. Cans enter the equipment on guide chains and are spun over the gas flames to raise the temperature to around 130°C in 30 s. The cans are then spray cooled before exiting the sterilisation system.

References

1. DHSS Code of Practice No. 10, *The Canning of Low Acid Foods*, HMSO, London, 1981.
2. Thorpe, R.H. and Atherton, D., Guidelines on the establishment of scheduled heat processes for low-acid foods, Technical Manual 3, Campden Food Preservation Research Association, Chipping Campden, Glos, UK, 1977.
3. Thorpe, R.H., Atherton, D. and Steele, D.S., Technical Manual 2, Campden Food Preservation Research Association, Chipping Campden, Glos, UK, 1975.
4. Tucker, G. and Clark, P., Computer modelling for the control of sterilisation processes, Technical Memorandum 529, Campden Food and Drink Research Association, Chipping Campden, Glos, UK, 1989.
5. Austin, G. and Atherton, D., Process control in hydrostatic cookers, Technical Manual No. 5, Campden Food Preservation Research Association, Chipping Campden, Glos, UK, 1975.
6. Bratt, L., May, N., Seager, A. and Williams, A., *Process control in reel and spiral cooker/coolers, good manufacturing practice guidelines*, Campden Food and Drink Research Association, Chipping Campden, Glos, UK, 1989.
7. Rose, D., Guidelines for the processing and aseptic packaging of low-acid foods, Technical Manual, 11, Campden Food Preservation Research Association, Chipping Campden, Glos, UK, 1986.
8. Richardson, P.S., and Gaze, J.E., Application of an alginate particle technique to the study of particle sterilisation under dynamic flow, Progress Report No. 6, Campden Food Preservation Research Association, Chipping Campden, Glos, UK, 1985.
9. Richardson, P.S. and Gaze, J.E., Application of an alginate particle technique to the study of particle sterilisation under dynamic flow, Technical Memorandum 429, Campden Food Preservation Research Association, Chipping Campden, Glos, UK, 1986.
10. *Food Manufacture* (May 1988).
11. Richardson, P.S., Flame sterilisation—a review, *J. Food Technol* 22 (1987) 3–14.

Further reading

Decareau R.V., *Microwaves in the Food Processing Industry*, Academic Press, New York, 1985.

4 Aseptic processing and packaging of heat preserved foods

K.E. STEVENSON and K.A. ITO

4.1 Introduction

Aseptic processing of foods has grown in starts and stops since its inception. Although we tend to think of this development as a rather recent one, in fact, in concept, it is over half a century old.

The heat-cool-fill process (HCF) was initiated by Ball in 1927 at the American Can Research Department. This process was designed to sterilise containers and product, bring them together aseptically and then to seal the containers such that the product would remain sterile. This was finally patented by Ball and put into commercial practice [1].

The aseptic process as we know it today, however, was initiated by Martin [2] with the development of the Dole Aseptic Process. This process is still utilised today. Commercial lines were installed producing soups, sauces, drinks and puddings. These products were in retail sizes and the product occupied a small but steady niche in the market place. This level of production continued until the late 1960s when canned pudding began to be a market factor [3]. Along with this product was a container innovation which revolutionised how the product was consumed; i.e. the two-piece container with the pull-tab lid, providing a convenience factor which helped usher aseptic processing to a new level of awareness.

About this time, the application of aseptic to bulk packaged items began. Mitchell [4] reports on the development in 1955 of 55-gallon drums of aseptically packed tomato paste. In the 1960s, the bag-in-box concept was introduced for aseptically packed dairy products. This was followed by the use of bag-in-box containers in large sizes for aseptically filled acid products. Dixon et al. [5] extended this concept to bulk tanks including tank cars. In 1976, the very large capacity tank concept was awarded the IFT industrial award [6]. Operations using large bulk containers continue today, and have facilitated year-round plant operations of seasonal products.

In the United States, another container-oriented advancement provided further impetus to the growth of aseptic packaged products. This occurred in January 1981 when the Food and Drug Administration [7] approved the use of hydrogen peroxide as a sterilising agent for polyethylene used in contact with food. Since that time aseptic packaging has grown to the level of over three billion packages [8] produced in 1989. With the use of a number of different flexible packaging materials has come the potential for varied presentation of products. These have included plastic single-serving cups of pudding, apple sauce and fruits, as well as various container types for individual servings of beverages and other products.

This growth and development has recently led the IFT [9] to name aseptic processing and packaging as the top invention in food science in the fifty years from 1939 to 1989. In order for this technology to fully realise its potential it will be necessary for all involved to ensure that products are produced in a maner which will provide a reliable, safe, quality product. This chapter discusses requirements for achieving this in aseptic processing and packaging systems.

4.1.1 Definitions

The following are definitions of terms used in this chapter.

Acid foods are foods that have a natural pH of 4.6 or below.

Acidified foods contain low acid food to which an acid or an acid food has been added to produce a food with a final equilibrium pH of 4.6 or less and a water activity greater that 0.85.

Aseptic describes a condition in which there is an absence of micro-organisms including viable spores. In the food industry the terms aseptic, sterile, and commercial sterility are often used interchangeably.

Aseptic system refers to the entire system necessary to produce a commercially sterile product contained in a hermetically sealed container. This term includes the product processing system and the filling and packaging system.

Aseptic processing system refers only to the system that processes the product and delivers it to the filling and packaging system.

Aseptic filling and packaging system refers to any piece of equipment that fills a sterile package or container with sterile product and hermetically seals it under aseptic conditions. These units or systems may also form and sterilise the package.

Aseptic zone is the area required to be made and maintained commercially sterile.

Commercial sterility of thermally processed food is the condition achieved by application of heat or other treatments to render the food free from micro-organisms of public health significance as well as those of non-health significance capable of reproducing under normal conditions of storage and distribution.

Commercial sterility of equipment and containers used for aseptic processing and packaging of food is the condition achieved and maintained by the application of heat or other methods which render such equipment and containers free from micro-organisms with public health significance as well as those of non-health significance capable of reproducing under normal conditions of storage and distribution.

Critical factor is any property, characteristic, condition, aspect, or other parameter which when varied may affect the scheduled process and the attainment of commercial sterility.

Hermetically sealed containers are containers designed to be secure against the entry of micro-organisms and to maintain commercial sterility of their contents.

Low acid foods are any foods with a finished equilibrium pH greater than 4.6 and a water activity greater than 0.85.

Water activity is a measure of the free moisture of a product. It is determined by dividing water vapour pressure of the substance by the vapour pressure of pure water at the same temperature.

4.2 Aseptic processing systems

The principles of heat preservation and heat processing equipment have been discussed in previous chapters but it is important to mention some unique aspects associated with aseptic processing systems. A general diagram of a system is presented in Figure 4.1. Common features of aseptic processing systems include:

- A metering device to control and document the flow rate of product through the system.
- A method of heating the product to sterilising temperatures.
- A method of holding the product at an elevated temperature for a time sufficient for sterilisation.
- A method of cooling to reduce the temperature of product prior to filling.
- A means to sterilise the system prior to production and to maintain sterility during production.
- Adequate safeguards to protect sterility and prevent non-sterile product from reaching the packaging equipment.

4.2.1 Flow control

The time required for sterilisation is directly related to the rate of flow of the fastest moving particle through the system. The fastest moving particle is in turn related to a function of the flow characteristics of the food. Thus, the

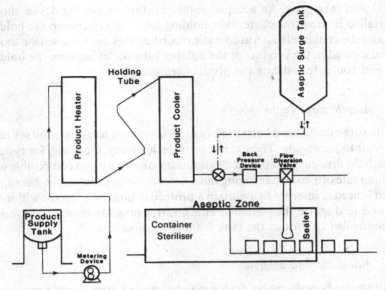

Figure 4.1 General diagram of an aseptic system.

aseptic processing system must be designed to assure product flow at a uniform rate. This is achieved by the use of a device called a timing or metering pump. Timing pumps may be variable speed or fixed rate. When a variable speed pump is used it must be protected against unauthorised changes in the pump speed which could affect the rate of product flow through the system.

4.2.2 Holding tube

Once the product has been brought to sterilisation temperature, it flows into a holding tube. The tube provides the required residence time at the sterilisation temperature. The process is designed to ensure that the fastest moving particle through the holding tube will receive a time/temperature process sufficient for sterilisation. Since the hold tube is essential for ensuring that the product is held at sterilisation temperatures for the proper time, certain precautions must be followed. The holding tube should slope upwards in the direction of product flow to assist in eliminating air pockets and prevent self-draining. Product in the hold tube must be maintained under pressure sufficient to avoid flashing or boiling since this can decrease the product residence time in the hold tube. This is usually accomplished by a back pressure device. Since there is some loss of temperature as product passes through the holding tube, the product temperature must be sufficiently high upon entering so that even with some temperature drop, it will still be above the prescribed minimum temperature at the exit of the holding tube. No external heating of the holding tube should take place. An acceptable temperature measuring device should be installed in the product steriliser holding tube outlet between the holding tube and the cooler inlet. An automatic recording thermometer sensor should also be located in the product at the holding tube outlet between the holding tube and cooler to indicate the product temperature.

4.2.3   Aseptic surge tanks

Aseptic surge tanks have been used in aseptic systems to allow product to be held prior to packaging. These tanks provide flexibility, especially for systems in which the flow rate of a product's sterilisation system is not compatible with the filling rate of a given packaging unit. A sterile air or other sterile gas supply system is needed in order to maintain a protective positive pressure within the tank and to displace the contents. This positive pressure must be monitored and controlled to protect the tank from contamination.

4.2.4   Automatic flow diversion

An automatic flow diversion device may be utilised in an aseptic processing system to prevent the possibility of unsterile product reaching the sterile packaging equipment. The flow diversion device must be designed so that it can be sterilised and operated reliably. Past experience has shown that a flow diversion valve of the gravity drain type should not be used in aseptic systems due to the possibility of recontamination of sterile product. The design and operation of a flow diversion system is critical. The flow diversion valve should divert product automatically if a deviation occurs. A few examples of situations which may cause a diversion are low temperature in the hold tube, inadequate pressure differential in regenerators, or below minimum operating specifications for the packaging unit.

4.2.5   Maintenance of sterility

Once the product leaves the hold tube it is sterile and subject to contamination if micro-organisms are permitted to enter the system. Product flow under pressure is one of the best ways to prevent contamination. The back pressure device is to prevent product from boiling or flashing and also helps maintain the entire product systems under elevated pressure. Effective barriers against micro-organisms must be provided at all potential sites of contamination. Examples include rotating or reciprocating shafts and stems of aseptic valves. Steam seals at these locations can provide an effective barrier but they must be monitored to ensure proper functioning. If other types of barriers are used there must be a means provided to permit the operator to monitor the proper functioning of the barrier.

4.3 Aseptic filling and packaging systems

4.3.1 General requirements

Aseptic filling and packaging systems must produce commercially sterile product in hermetically sealed containers. In order to achieve this result, an aseptic filling and packaging system is designed to:

- Sterilise equipment surfaces in order to create an aseptic zone
- Sterilise the food contact surfaces of the packaging material
- Aseptically fill product into the container in an aseptic zone
- Hermetically seal containers
- Maintain sterility during production

Aseptic filling and packaging systems for acid or acidified foods may be relatively simple. In contrast, systems for use with low acid foods are normally complex due to requirements for monitoring and controlling factors which assure that products are safe. Except where noted, the following sections discuss systems employed for producing low acid foods.

4.3.2 Sterilisation agents

Heat, chemicals and radiation have been used, alone or in combination, for sterilisation of aseptic equipment and packaging materials. Practical considerations and regulatory requirements have limited the number of sterilants which are used for aseptic systems.

Heat Initially, heat was used as the sterilant for aseptic systems as a natural extension of thermal processing. Product supply lines and fillers are commonly sterilised by 'moist' heat in the form of hot water or saturated steam under pressure. 'Dry' heat, in the form of superheated steam or hot air, may also be used to sterilise equipment. However, due to the relatively high dry heat resistance of bacterial endospores, the time-temperature requirements for dry heat sterilisation are considerably higher than those for moist heat sterilisation.

Since relatively large masses of metal are often present in aseptic filling and packaging systems, high temperatues and relatively long holding periods are necessary to assure that appropriate sterilisation has occurred. Systems employing moist heat are frequently sterilised at temperatures ranging from 121–129°C, while 176–232°C is used for sterilisation by dry heat. In addition, sterilisation of air by incineration usually is conducted at temperatures ranging from 260 to 315°C.

Chemicals Hydrogen peroxide is the overwhelming choice for use as a chemical sterilant. Other chemicals which have been used as sterilants, primarily for use in systems for acid foods, include various acids, ethanol, ethylene oxide and peracetic acid.

In the United States, chemical sterilants cannot be used for sterilisation of packaging materials unless a specific food additive regulation permits use. For example, while hydrogen peroxide had been used in European aseptic systems since the 1960s and hydrogen peroxide was used in the United States in certain foods as a bleaching agent and antimicrobial agent, it was not approved for use in the United States as a sterilant for packaging material until 1981. In addition to requiring the use of 35% hydrogen peroxide only on polyethylene food contact surfaces, the initial food additive regulation included a requirement to assay for hydrogen peroxide residuals. The regulation required that packages be filled with distilled water (under normal operating conditions) and that the residual hydrogen peroxide in the water should be no more than 0.1 ppm. The food additive petition has been modified over the years to allow use of a variety of packaging materials and to increase the allowable hydrogen peroxide residual to 0.5 ppm.

Hydrogen peroxide is not an efficient sporicide when used at room temperature. However, the sporicidal activity increases substantially with increasing temperatures. Thus, most aseptic filling and packaging systems utilise hydrogen peroxide in combination with heat for rapid sterilisation of equipment and/or packaging material. Critical factors associated with use of hydrogen peroxide are discussed in a separate section.

It is doubtful whether there will be widespread use of chemical sterilants other than hydrogen peroxide in the foreseeable future. This is a result of the extensive data required concerning the potential toxicity of compounds, as well as the need for practical and reliable procedures for monitoring concentrations of sterilants and detecting their residuals.

Radiation   Gamma-radiation has been used for decades to decontaminate packaging materials for use in aseptic systems for packing acid and acidified foods. Due to the penetrating powers of gamma-radiation, pouches and bags have been treated in bulk at commercial irradiators. A dose of approximately 1.5 Megaradians (Mrad) is commonly used to decontaminate containers for acid and acidified foods. Recently, the FDA has accepted processes for two low acid food aseptic filling and packaging systems which utilise bags presterilised by radiation. Doses required to sterilise containers for use with low acid foods are considerably higher than those required for acid and acidified foods.

Other types of radiation are not widely used in aseptic systems. Ultraviolet (UV-C) light has been used to decontaminate food contact surfaces. The low penetration and problems associated with 'shadowing' will limit the use of UV-C for aseptic systems packing low acid foods. While equipment size, speed and costs have precluded use of 'electron beam' irradiators up to the present, it is only a matter of time before such a system is developed.

Table 4.1 Classification of aseptic filling and packaging systems.

Category	Examples of systems
I. Metal and rigid containers sterilised by heat	
A. Steam/metal containers Imdec	Dole canning systems Drum fillers, e.g. Scholle, FranRica,
B. Hot air/composite can	Dole hot air system
II. Webfed paperboard sterilised by H_2O_2	Tetra Pak (Brik Pak) International Paper
III. Preformed paperboard containers	Combibloc LiquiPak
IV. Performed, rigid/plastic containers	Metalbox Freshfill Gasti Crosscheck
V. Thermoform-fill-seal	Benco Asepack Bosch Servac Conoffast Thermoforming USA
VI. Flexible plastic containers	
A. Bag-in-box type	Scholle LiquiBox
B. Pouches	Asepak Prepac Prodo Pak Inpaco
C. Blow moulded	Bottlepack Serac ALP

Adapted from Stevenson [30].

4.3.3 Classification of aseptic packaging systems

Aseptic filling and packaging systems can be classified into categories based on the type of packaging material and the method of forming the container (Table 4.1).

Rigid containers sterilised by heat The Dole canning systems which use superheated steam to sterilise metal container bodies and lids, are the only systems for use with low acid foods. Several systems, including drum fillers are used for acid and acidified foods.

Webfed paperboard sterilised by hydrogen peroxide Several models of Tetra Pak (Brik Pak) aseptic filling and packaging systems are used throughout the world for packing low acid foods. The International Paper units for packing low acid foods have not been widely marketed. Both systems utilise hydrogen peroxide and heat to sterilise a web of packaging material prior to forming and sealing the containers (Figure 4.2).

Figure 4.2 Tetra Pak AB9.

Preformed paperboard containers The Combibloc systems preform cartons prior to sterilisation using hydrogen peroxide and heat. Other systems have been used for packing low acid foods in Europe (Figure 4.3).

Preformed rigid plastic containers The Metal Box Freshfill aseptic filling and packaging systems utilise hydrogen peroxide and heat to sterilise preformed cups. Other systems in use in Europe include Gasti and Hamba. Systems using UV-C light or saturated steam for package sterilisation are in the developmental stages (Figure 4.4).

Thermoform-fill-seal Bosch thermoforming aseptic systems and privately modified Benco Asepack systems utilise hydrogen peroxide and heat to sterilise the packaging materials. The Conoffast unit employs a unique method

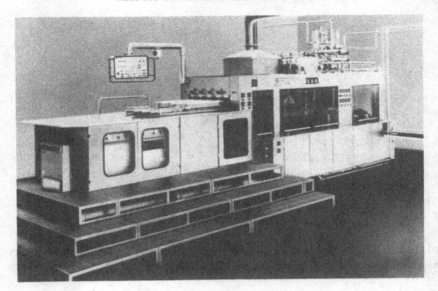

Figure 4.3 Combibloc CF716A.

Figure 4.4 Metalbox Freshfill ML-4.

Figure 4.5   Bosch Servac 78AS.

of sterilisation by the heat of co-extrusion and use of a peelable layer in its aseptic filling and packaging system (Figure 4.5).

Flexible plastic containers   The Scholle and FranRica systems both use flexible plastic containers which have been presterilised via gamma-radiation. The former uses hydrogen peroxide and heat, while the latter uses saturated steam, to sterilise a minichamber and the exterior of the cap area to facilitate the aseptic filling operation (Figure 4.6). In addition, Bosch is developing an aseptic system which sterilises a plastic web, using hydrogen peroxide and heat, prior to forming pouches. In addition, one blow-moulding aseptic

Figure 4.6   Bosch TFA 4940.

system, a modified ALP unit, has been accepted for packing low acid foods in the United States.

4.4 Sterilisation procedures

4.4.1 Sterilisation of processing equipment

The aseptic processing system must be brought to a condition of commercial sterility prior to production of sterile product. Sterilisation of equipment food contact surfaces is usually accomplished by circulation of hot water through the system for a length of time sufficient to destroy organisms which might be present. However, in some systems, or at least portions of some systems, saturated steam is used. The whole system must be maintained at or above a specified temperature by recirculating a heating medium for a required period of time. This includes the holding tube, the product coolers and, where utilised, aseptic pumps, aseptic surge tanks, and the sterile side of product-to-product regenerators and all connecting piping and valves. Due to their large volume, aseptic surge tanks and flash chambers are commonly sterilised with saturated steam.

The equivalent of 30 min at 121°C is normally the minimum recommended for sterilisation of equipment for low acid aseptic processing systems. For acid or acidified aseptic operations it is possible to successfully sterilise with a lower time/temperature combination; 30 min at 104°C is a frequently used minimum value. Acidification of the presterilisation water to pH 3.5 or below is a common practice which enhances the effectiveness of the presterilisation cycle for acid products.

In order to properly control presterilisation of equipment, thermometers or thermocouple devices should be used to determine when various points in the system achieve the required sterilisation temperature. Temperature recording devices should also be used to provide a record that the proper presterilisation cycle was applied before each production run.

4.4.2 Sterilisation of products

The principles involved in thermal sterilisation of foods remain the same whether one is attempting sterilisation of products in containers or sterilisation of products prior to filling in the final container. Thus, it is necessary to know the thermal destruction rate for the micro-organisms of consequence in the food being processed. The procedures necessary for acquisition of such data are available [10]. Then, it is necessary to apply this information properly so that the appropriate time and temperature for destruction of the organisms can be achieved.

Sterilisation procedures for products in containers usually require long sterilisation times, since the heat transfer to the product is relatively slow. Sterilisation prior to filling in the container, as accomplished in aseptic processing, requires relatively short heating periods. This sterilisation process is usually accomplished by heating product rapidly to 130–145°C, holding for an appropriate time, then rapidly cooling the product. The specific product will determine the actual combination of temperature and time required for sterilisation.

The product heating is accomplished by either indirect or direct heat exchange. In liquid homogeneous products, indirect heating occurs when the product to be heated and the heating medium are kept separated by the heating surface. This type of heating is accomplished using a scraped surface, tubular or plate heat exchanger. Direct heat exchange is achieved by placing the heating medium directly into the product. This is accomplished by injection of steam into the product or vice versa. In either case the added water from the steam must either be removed or accounted for in the formulation. The specific type of heating used is usually dictated by the nature of the product and the economics of operation.

Following heating to the sterilising temperature, the product must be held at this temperature for a sufficient length of time to accomplish sterilisation. The sterilisation time in continuous flow aseptic systems is obtained by having product flow through a non-heating pipe attached to the heating systems. This pipe is called the holding tube. This tube is of a specified uniform diameter and length. Its capacity is such that the fastest particle is held at the minimum time required to sterilise the product. The time which the product remains in the tube is dependent not only upon the holding capacity of the tube and the rate at which the product is pumped through the tube, but also upon the manner in which the product flows through the tube.

Product flow through round pipes can generally be considered to occur either as plug, turbulent or laminar flow. In plug flow, all components flow through the tube at the same rate as a uniform mass. Thus, the flow velocity of all components is the same. This rarely occurs. In fully developed laminar flow, the product flows slowly next to the pipe walls and fastest at the centre line. The flow velocity appears parabolic with the fastest component at the centre line theoretically moving at twice the average rate. In turbulent flow the fastest moving component is also at the centre line. Although the relationship of this fastest component to the average varies, it is usually assumed to be about 1.25 times the average flow rate. The calculation of a dimensionless term, the Reynolds number is used to determine the type of flow. The Reynolds number may be calculated using the formula [11]

$$Re = \frac{6.31\,w}{d\mu}$$

where Re = Reynolds number, w = mass flow (1b/h), d = pipe diameter (inches), μ = fluid viscosity (cP). The flow will generally be laminar if the

Reynolds number is less than 2000. When the flow type has not been determined, the conservative assumption of fully developed laminar flow is usually used to determine residence time.

Residence time or its corollary, holding tube length, can be calculated assuming fully developed laminar flow using the formula [12]

$$L = \frac{8Qt}{D^2}$$

where L = length of the holding tube (inches), Q = volumetric flow through the hold tube (gal/min), t = hold time (minutes), D = inside diameter of hold tube (inches). Thus by knowing the death rate of the significant organism in the product, it is possible to determine the time the product should be held in the holding tube at a processing temperature that will sterilise the product.

Processing of homogeneous fluid products is usually not difficult to accomplish once the flow characteristics have been determined and the proper hold time has been established. The fluid products currently processed in aseptic systems fall into two categories. Acid products are those with pH values $\leqslant 4.6$, and low acid products those with pH values greater than 4.6. In the acid category are many juices, drinks and fruit products. Citrus juices [13], papaya and guava purée [14] are some examples of acid products which are currently processed. Although micro-organisms must be destroyed to sterilise the product, the shelf-life and product quality are often dependent upon retention of nutrients and the destruction of relatively heat resistant enzymes. Marcy [13] reported that the destruction of the pectinesterase enzyme in citrus juices is necessary to produce a stable product. Since this enzyme is more heat resistant than the micro-organisms usually associated with product spoilage, destruction of pectinesterase during processing of these products (87–100°C) assures sterility of the products as well as their stability.

The low acid food category includes many different kinds of products such as custards, puddings, fluid milk and various sauces. These products must receive a process which is sufficient to ensure the destruction of *Clostridium botulinum* spores. Then, to ensure commercial sterility, it may be necessary to provide a more stringent process to destroy more resistant spore-formers which could cause spoilage. The destruction of these micro-organisms will occur at the temperatures usually used for these products (130–150°C). Again consideration of the resistance of nutrients and enzymes is necessary.

In milk it is necessary to destroy micro-organisms to prevent spoilage and to assure public health safety. At the temperatures usually used for treatment of this product (135–150°C), this is accomplished rather easily [15]. However, just as in the acid foods, nutrient retention may be significantly improved whereas failure to destroy enzyme systems [16] could affect the stability of the product.

Thus, for low acid products it is necessary to first consider the destruction of micro-organisms to ensure sterility and public health safety, but it is also necessary to look at other factors such as nutrient retention and enzyme destruction to ensure that a saleable, stable product will be produced.

Stringent regulatory requirements in the United States have hampered the commercial application of aseptic processing and packaging of products containing discrete particles. The US regulations require that processes for low acid foods be sufficient to protect the public health and produce a commercially sterile product. Recently, the FDA specified several factors which must be considered in the development of thermal processes for products [17]. The factors included:

- Selecting an appropriate sterilising value (F)
- Predicting the sterilising value delivered by the thermal process utilising the following:
  - mathematical model
  - particle residence time
  - particle to fluid heat transfer coefficient
- Biological validation of thermal process recommendations.
- Specifications of critical factors

Adequate processing and packaging equipment is available for commercial production of products containing discrete particles [18]. While products containing discrete particles have been produced in Europe, there is a lag in the implementation of this technology in the United States. In the absence of appropriate experimental data, Dignan *et al.* [17] recommended that the most conservative approach be taken at each stage. This results in what many consider to be an overly conservative approach to processing of such products. In addition, these regulators have considered that a given process would be specific to the product and the processing system. Thus, extensive experiments would be required in order to accumulate appropriate experimental data and would need to be repeated for duplicate processing systems or slight changes in product formula.

A variety of techniques have been used for mathematical modelling of aseptic processes in order to predict sterilising values. Several such procedures have been reported in the literature [19–23]. Particle size and proportion, convective heat transfer coefficient at the particle/fluid interface, and residence time distribution within the system, in addition to system configuration and product physical and thermal properties are all important factors which must be considered in process calculations. Alternatively, if experiments are conducted to determine the heating characteristics of particles within the product then this information can be used for process development. Mathematical modelling will allow researchers to artificially manipulate various operating factors and assess the impact of various changes in operating parameters.

One of the major concerns at the present time is the ability to address the residence time of the particles during the process. Particle residence time will be influenced by particle/fluid interaction as well as other parameters such as rheology, flow rate, system configuration, and in the case of scraped surface heat exchangers, the mutator rotation velocity. Currently methodology is available for the determination of particle residence time distribution in a hold tube. Further improvements should also be expected as better instrumentation becomes available. McCoy et al. [24] studied the problem of particle residence time in a hold tube and indicated that the particle velocity could exceed the average fluid velocity. However, this type of analysis could be misleading since it would be more appropriate to compare the fastest particle velocity to the fastest fluid velocity or mean particle residence time to mean fluid residence time.

The rate at which heat can be transferred from the fluid to the particle is limited by the convective heat transfer coefficient at the particle/fluid interface [17, 20, 21, 25, 26]. The convective heat transfer coefficient at the particle/fluid interface is a function of relative motion between the particle and the fluid, the rheological properties of the fluid, and the shape and size of the particles. For accurate process calculations, the heat transfer coefficient must be determined experimentally for the product in question, and sufficient information must be available to permit the use of a conservative estimate.

Biological methods can be used for process determination and/or process validation. For process determination a count reduction procedure can be accomplished using inoculated particles with a known initial inoculum level. The particles must be recovered at the end of the process and this may require an extremely large number of experiments to obtain useful data. Biological validation using the inoculated pack technique is often used to verify the ability of a processing system to deliver a predetermined lethality [27, 28]. In this procedure product inoculated with an appropriate test organism is processed per the mathematically determined process and at several lower temperature levels. Product samples are collected and checked for surviving micro-organisms. In an aseptic processing system, the results of an inoculated pack will reflect lethality delivered in the whole system including the heater, hold tube and in the coolers.

4.4.3 Sterilisation of filling and packaging equipment

Prior to producing any product, the aseptic and packaging equipment food contact surfaces must be brought to a condition of commercial sterility. This is generally done in two broad categories: (a) machine or equipment sterilisation and (b) sterilisation of the air supply system. The filling area is normally sterilised with the aseptic processing system. Other areas and equipment surfaces must be sterilised in order to create an aseptic zone. The aseptic zone is the area within the aseptic packaging machine which is sterilised and maintained sterile during production. The aseptic zone is considered to begin where the packaging material is sterilised or where the presterilised packaging material is introduced into the machine. The aseptic zone ends after the packages are sealed to protect them from contamination by micro-organisms. All areas between these two points are considered as part of the aseptic zone. Since the area in the aseptic zone may contain a variety of surfaces, including moving parts composed of different materials, sterilants must be uniformly effective and their application controlled throughout the entire zone.

Sterilisation of the air supply system includes sterilisation of the air filters and the air supply lines. In many systems a large supply of sterile air is required during operation. Thus, initial sterilisation of the filters and lines is of paramount importance. One advantage of incineration of air supplies is the ability to continuously monitor the temperature of incineration. Thus, there is a record of the ability or inability to sterilise the air supply. While filtration can be used successfully to sterilise air supplies and while there are tests to detect gross failures in filters, it is difficult to utilise any on-line tests which can guarantee that a filtration system is providing sterile air. Whether incineration or filtration is used, appropriate monitors must be included to assure proper sterilisation of the air supply system and also maintenance of a sterile air supply.

Once the aseptic filling and packaging unit has been sterilised, sterility must be maintained during production. The aseptic zone should be constructed in a manner which provides sterilisable physical barriers between sterile and non-sterile areas. Mechanisms must be provided to allow sterile packaging materials and hermetically sealed packages to enter and leave the aseptic zone without compromising the sterility of the area. The sterility of the aseptic zone can be protected from contamination by maintaining the aseptic zone under positive pressure of sterile air or other gas. As finished containers leave the sterile area, sterile air flows outward preventing contamination from entering the aseptic area. The sterile air pressure within the aseptic zone must be kept at a level proven to maintain sterility of the zone.

4.5 Ensuring product safety

The concept of critical factors has been used to help ensure that thermally processed products are safe. This concept can be used for processing operations as well as filling and packaging operations. In general terms, we must accomplish the following: (a) determine which factors are critical to a given operation and once these factors are known, (b) set minimum or maximum values so that the critical factors are controlled within appropriate boundaries. In addition, it is necessary to (c) monitor the critical factors and (d) design appropriate alarms, warning devices or 'stops' which will be included in the aseptic system operations.

To illustrate establishing critical factors, sterilisation of packaging material by hydrogen peroxide is used as an example. In order to use this sterilant, the US regulations state that food contact surfaces must be sterilised and there must be a residual of 0.5 ppm or less. The concentrations of hydrogen peroxide are normally 30–35% for most systems. Since the sporicidal rate of hydrogen peroxide is dependent in part on concentration, a minimum limit must be set. Furthermore, since higher concentrations could lead to unacceptable residuals, a maximum limit must also be set. Similar reasons require minimum and maximum values to be set on the amount of hydrogen peroxide that is applied to the container. Other factors associated with application will also be critical to obtaining appropriate destruction of micro-organisms or hydrogen peroxide residuals. For systems which employ hydrogen peroxide baths, in addition to measuring the concentration of hydrogen peroxide, the operator must ensure that a surfactant or wetting agent is present if needed, the consumption of peroxide is monitored, the bath must contain an appropriate level of peroxide, the machine must monitor wetting of the surface of packaging material, the heating element and air temperatures must meet minimum values and an appropriate positive pressure of sterile air must be maintained. All of these operations are critical for microbial destruction and/or acceptable residuals. A similar list of factors can be made for systems which apply a spray of hydrogen peroxide. A surfactant may or may not be required. Other factors which should be monitored include peroxide level in the supply tank, air pressure or flow to facilitate spraying, the amount of spray, the height of the spraying nozzle and angle of the spray, and most importantly, the fact that each package is sprayed and that the nozzle is not clogged. Other operations such as air flow during drying and times and temperatures for application and drying are critical. These items note some of the complexities of identifying and establishing critical factors. Furthermore, extensive microbiological tests are required to establish appropriate limits for these critical factors. Then a monitoring system must be put in place to ensure that proper sterilisation takes place.

The National Food Processors Association [29] has established the following criteria for aseptic systems for packaging low acid foods:

(1) Sufficient data must be provided to show that the equipment has been tested and the methods utilised are adequate to ensure and maintain sterility. Bacteriological testing procedures and results must be provided.

(2) Plans or drawings of the product sterilising system must be submitted, showing the location of monitoring devices, such as thermometers, or thermocouple devices for temperature measurement.

(3) If a chemical sterilant is intended for sterilisation of containers, information must be provided to show the procedure for maintaining an adequate and constant concentration of the sterilant, including test procedures for measuring sterilant concentration.

(4) If the sporicidal effect of the chemical, as used in the system, is dependent on temperature and/or time or any other physical condition, information must be provided to show how this is controlled and what system of alarms will signal when conditions are out of control.

(5) Sufficient data must be provided to show that non-thermal methods of container sterilisation do not enable a residue of the chemical sterilant to be carried through and come in contact with the food product

(6) In addition, data must be provided to ensure that the procedures used do not cause formation of harmful substances within the packaging material as a result of its exposure to the sterilising procedure.

(7) Information must be provided to demonstrate the integrity of filled and closed containers under normal and intensified conditions of commercial warehousing and transportation.

(8) Information must be provided regarding procedures for on-line monitoring of the adequacy of the container closure to ensure against the entrance of micro-organisms.

(9) Finally, it must be understood that each individual commercial installation for packaging low acid foods must undergo on-site testing to demonstrate the adequacy of the system to comply with the conditions outlined above.

References

1. Ball, C.O. and Olson, F.C.W., McGraw-Hill, New York, 1957, p. 96.
2. Martin, W.M., *Food Ind.* 20 (1948) 1069.
3. Hilliker, F., *Canner/Packer* 13 (1968) 17.
4. Mitchell, E.L., *Adv. Food Res.*, 32 (1988) 2.
5. Dixon, M.S., Marshall, R.B., and Crerar, J.B., U.S. Patent No. 3,096,161, 1963.
6. Anonymous, *Food Technol.* 30 (1976) 23.
7. Food and Drug Administration, Federal Register 46 No. 6 (1981) 2341.
8. Brody, A.L., *Prepared Foods* 15 (1989) 166.
9. Anonymous, *Food Technol.* 43 (1989) 309.
10. Stumbo, C.R., *Thermobacteriology in Food Processing*, 2nd edn., Academic Press, New York, 1973.
11. Ludwig, E.E., *Chem. Eng.* 67 (1960) 161.
12. Dickerson, Jr., R.W., Scalzo, A.M., Reed, Jr., R.B. and Parker, R.W., *J. Dairy Sci.* 5 (1968) 1731.
13. Marcy, J.E., *Transactions of the 1982 Citrus Engineering Conference*, ASME, Lakeland, Florida, Vol. 28 (1982) 18.
14. Chan, Jr., H.T. and Cavaletto, C.G., *J. Food Sci.* 47 (1982) 1164.
15. Mehta, R.S., *J. Food Protect.* 43 (1980) 212.
16. Adams, D.M., Barach, J.T. and Speck, M.L., *J. Dairy Sci.* 58 (1975) 828.
17. Dignan, D.M., Berry, M.R., Pflug, I.J. and Gardine, T.D., *Food Technol.* 43 (1989) 118.
18. Wernimont, D.V., *Proceedings of National Food Processors Association Conference— Capitalizing on Aseptic II*, Food Processors Institute, Washington, DC, 1985 p. 59.
19. de Ruyter, P.W. and Brunet, R., *Food Technol.* 27 (1973) 44.
20. Manson, J.E. and Cullen, J.F., *J. Food Sci.* 39 (1988) 1084.

21. Sastry, S.K., *J. Food Sci.* 51 (1986) 1323.
22. Chandarana, D.I., Gavin, III, A. and Wheaton, F.W., Mathematical model for aseptic processing of particulates in liquid, presented at the Institute of Food Technologists, Annual Meeting, New Orleans, LA, June 19–22, 1988.
23. Chandarana, D.I. and Gavin, III. A. *J. Food Sci.* 54 (1989) 198.
24. McCoy *et al.* (1987).
25. Chandarana, D.I. Unpublished Ph.D. dissertation, University of Maryland, College Park, MD, 1988.
26. Chandarana, D.I., Gavin, III. A. and Wheaton, F.W., Particle fluid interface heat transfer during aseptic processing of foods, Paper 88-6599, American Society of Agricultural Engineers, St. Joseph, MI, 1988.
27. National Canners Association, *Laboratory Manual for Food Canners and Processes*, AVI, Westport, CT, 1968.
28. Bernard, D.T., Gavin, A., Scott, V.N., Polvino, D.A. and Chandarana, D., *Principles of Aseptic Processing and Packaging*, eds. Nelson *et al.*, The Food Processors Institute, Washington, DC 1987.
29. Stevenson, K.E., *Proceedings of National Food Processors Association Conference— Capitalizing on Aseptic*, Food Processors Institute, Washington, DC, 1983, p. 21.
30. Stevenson, K.E., *Proceedings of National Food Processors Association Conference— Capitalizing on Aseptic II*, Food Processors Institute, Washington, DC, 1985, p. 59.

5 Packaging of heat preserved foods in metal containers

T.A. TURNER

5.1 Introduction

Metal cans have dominated sectors of the food and beverage markets for many years because of their cost effectiveness, durability and the overall protection they provide for their contents. The past 20 years have seen dramatic changes in can-making technology and in the materials used to make cans and closures. Can manufacture has become increasingly a high technology business.

Food cans can be made from steel, in a variety of forms, and from several aluminium alloys, using a number of different manufacturing routes. This chapter reviews the various can-making methods, the metals used in can manufacture and the organic coating systems used in their protection, all of which will be seen to be closely inter-related.

5.2 Metals used in can manufacture

Containers for heat processed foods are made from steel, in a variety of forms, and from aluminium. The former is substantially more common than the latter for reasons of cost and performance.

5.2.1 Steel

Steel, usually in the form of tinplate (hence the misnomer, tin can), is by far the most common metal used in the manufacture of heat processed food cans. Over 50 billion such cans are made each year. The gauge and, where tinplate is used, the level of tincoating varies considerably with container size and the product to be packed. Typical ranges are:

- Nominal gauge, 0.15–0.30 mm
- Tincoating weight, 0.5–15 gsm

The generic name for steel-based materials is tin-mill products, a name derived from the equipment used in their production. In fact tinplate and various tin-free steels are produced on essentially the same equipment.

Tin-mill products are available in a wide range of specifications relating to

- Gauge
- Single or double reduction
- Temper
- Continuous or batch annealed (CA or BA)
- Continuous or ingot cast

A specification is selected to suit the product to be packed and the manufacturing route for the container.

The manufacture of steel-based strip products is shown in simplified form in Figure 5.1. The essential steps are seen to be as follows:

(1) Refined low carbon steel, a typical composition for which is shown in Table 5.1, is cast to ingot or continuously cast into a slab. In practice, the length of the slab is limited only by the supply of molten steel from the furnace tundish.

(2) In the case of ingot cast steel, the ingot is reheated to uniform temperature and then rolled into a slab; once in slab form the steel is hot rolled into a strip.

(3) In strip form, the steel has a significantly thick iron oxide 'scale' which is removed by mechanical flexing of the strip followed by passage through acid tanks (commonly sulphuric or hydrochloric acid). This process is called 'pickling' and is followed by oiling to facilitate subsequent rolling.

(4) The next stage involves cold reduction to nominal gauge, which in the case of single reduced plate is close to the final gauge. Typical cold reduction can result in a 10-fold reduction in thickness.

(5) Cold reduction produces a material which is fully hard and virtually unusable by virtue of its crystalline structure which renders it very brittle.

(6) The next important stage is annealing which traditionally necessitated holding the coils of steel at elevated temperatures (around 580–600°C) in an inert atmosphere for a prescribed length of time (typically 60 h incorporating 8–10 h at peak temperature). More recently continuous annealing that allows less time for crystal growth, has been introduced, this can be be carried out in the coil as a continuous process in a matter of minutes.

(7) The two processes of annealing produce different products. Batch annealing produces steel within the temper range 1–3 whilst continuous annealing produces higher tempers of 4–6. In each case the precise level is dictated by chemistry. The annealed product can then either be light-

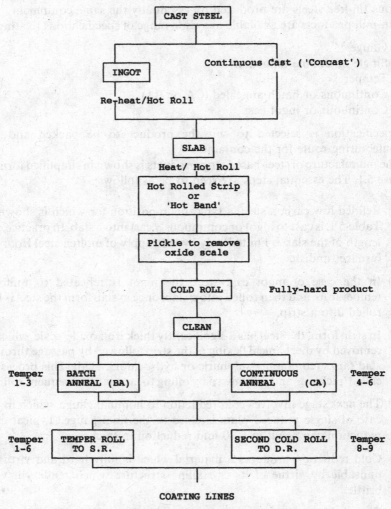

Figure 5.1 The manufacture of tin-mill products.

temper rolled to produce the surface finish required or be subjected to further cold reduction to produce DR materials of temper 8 or 9. The latter processes produce the final gauge.

(8) At this stage the strip is ready for final finishing:

    – oiling in the case of blackplate
    – electrolytic tinning in the case of tinplate
    – electrolytic deposition of a chrome/chromium oxide layer in the case of electrolytic chrome coated steel (ECCS), commonly referred to as tin-free steel (TFS)

Table 5.1 Steel grade CP (low copper content).

	% max
Carbon	0.13
Manganese	0.60
Phosphorus	0.02
Sulphur	0.05
Silicon	0.02
Copper	0.08
Aluminium	0.08
Nitrogen	0.01
Plus traces of arsenic and nickel	

Table 5.2 Coating operations[a].

Blackplate	Tinplate	ECCS (TFS)
Oil	Clean	Clean
	Electroplate	Electrolytic deposition
	Flow brighten	of chromium
	(heat to melting	Electrolytic deposition
	point and quench)	of chromium oxide
	Chromate passivation	Oil
	(300/311)	
	Oil	

[a] Oil may be butyl stearate, dioctyl sebacate or acetyl tributyl citrate.

A summary of these coating processes is given in Table 5.2. The above is intended to give the reader a broad understanding of the steel-making process and a familiarity with common terminology used in the can-making industry. More information on steel and strip manufacture and on the metallurgical properties produced by the various manufacturing routes can be found elsewhere [1–3].

5.2.2 *Tin-free steels (TFS) and blackplate*

Blackplate, defined as uncoated mild steel, has been considered for the manufacture of food cans but is likely to be suitable only for a very limited range of products even when fully lacquered. This is because it readily rusts and generally has poor chemical resistance.

Tin-free steel (ECCS) has found fairly wide usage, typical examples being draw-redraw containers and fixed (non-easy open) ends for processed food cans. The extremely abrasive surface of TFS material necessitates overall lacquering prior to fabrication of containers or components to avoid tool wear. This aspect is dealt with later.

Table 5.3 Composition of a number of aluminium alloys.

Alloy type	Use	Added % (range)		Added % (max)[a]					
		Mn	Mg	Si	Fe	Cu	Cr	Zn	Ti
3004	Body stock	1.0–1.5	0.8–1.3	0.3	0.7	0.25	–	0.25	–
5182	End stock	0.2–0.5	4.0–5.0	0.2	0.35	0.06	0.1	0.25	0.1
5052		0.1 (max)	2.2–2.8	0.45	0.45	0.1	0.15–0.35	0.1	–
5042	Tab stock	0.2–0.5	3.0–4.0	0.2	0.35	0.15	0.1	0.25	0.1
5082		0.15	4.0–5.0	0.2	0.35	0.15	0.15	0.25	0.1

[a]Unless indicated otherwise, other ingredients up to 0.05% per element/total 0.15%.

In addition to differences in chemical composition, steel-based products are used in a wide range of strengths (tempers) and ductility, these differences arising from the detailed chemistry and from the particular choice of manufacturing route (see below).

In an increasingly high technology business, the importance of these production variables and the resultant variation in the characteristics of the steel, especially strength and plastic anisotropy can be of vital importance.

5.2.3 Aluminium

Aluminium in can-making gauges is used less extensively than steel for food can manufacture, although of course aluminium foils are used extensively. Today, commercial applications largely concentrate upon shallow drawn containers for such products as paté and fish. A typical manufacturing route for aluminium alloys, the composition of which is given in Table 5.3 is shown in Figure 5.2.

Aluminium in foil gauges (< 0.1 mm) is used extensively in trays for a variety of products including ready-meals, snacks and reheatable trays.

5.2.4 Mechanical properties

Mechanical properties of the various metals described above are important in the contexts of both container fabrication and the container's strength necessary to satisfactorily withstand filling/closure, retorting, and distribution through the retail chain.

After conversion into containers, aluminium alloys can achieve an ultimate strength comparable to that of the lowest temper steel. However, steel and the commonly used aluminium alloys are very different in terms of strength, strain hardening, tensile elongation and the reaction to exposure to lacquer curing temperatures (200°C approx.). Typical properties for 3004 H19 aluminium alloy and steel in three tempers are summarised in Table 5.4.

In the manufacture of draw-redraw (DRD) and drawn and wall-ironed (DWI) containers, grain size and plastic anisotropy are also important. Planar anisotropy, the asymmetric reaction to deformation during drawing or stamping associated with rolling direction and grain orientation, manifests itself as 'earing' (see Figure 5.3) and needs to be kept to a minimum to avoid excessive material wastage due to trimming.

The degree of plastic anisotropy is highly dependent upon chemical composition, in particular aluminium and nitrogen content, hot rolling conditions, the degree of primary (and secondary) cold reduction and the method of annealing.

The following properties are measured as routine by steel and can makers:

- Strength and tensile elongation by tensometer
- Plastic anisotropy

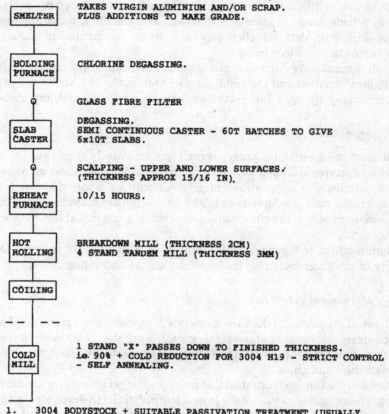

SMELTER — TAKES VIRGIN ALUMINIUM AND/OR SCRAP. PLUS ADDITIONS TO MAKE GRADE.

HOLDING FURNACE — CHLORINE DEGASSING.

GLASS FIBRE FILTER

SLAB CASTER — DEGASSING. SEMI CONTINUOUS CASTER - 60T BATCHES TO GIVE 6x10T SLABS.

SCALPING - UPPER AND LOWER SURFACES. (THICKNESS APPROX 15/16 IN).

REHEAT FURNACE — 10/15 HOURS.

HOT ROLLING — BREAKDOWN MILL (THICKNESS 2CM) 4 STAND TANDEM MILL (THICKNESS 3MM)

COILING

COLD MILL — 1 STAND "X" PASSES DOWN TO FINISHED THICKNESS. i.e. 90% + COLD REDUCTION FOR 3004 H19 - STRICT CONTROL - SELF ANNEALING.

1.   3004 BODYSTOCK + SUITABLE PASSIVATION TREATMENT (USUALLY BASED UPON CHROMATE)
2.   TAB STOCK DECREASED + DOS

3.   END STOCK 5182 DEGREASER CHROME/PHOS SURFACE TREATMENT + LEVELLER.

VARIANTS - THICKNESS AFTER HOT ROLLING. APPROX 6mm - INTERMEDIATE ANNEALING (BATCH ANNEAL) UNDERTAKEN DURING COLD ROLLING SCHEDULE.

Figure 5.2 Typical manufacturing route for aluminium container stock.

Table 5.4 Mechanical properties of steel and H19 aluminium alloys.

		3004 HI9	T2BA	T4CA	DR8BA
0.2% proof stress } N/mm²		285	235	315	550
UTS		295	350	400	580
Tensile elongation (%)		3	35	25	2
Strain hardening coefficient		0.04	0.18	0.15	0.04

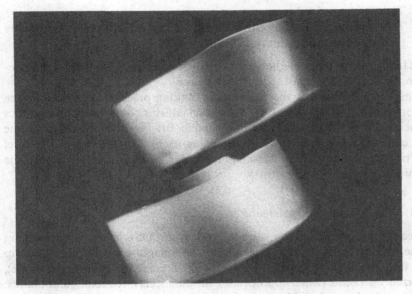

Figure 5.3 Drawn cups showing earing due to planar anisotropy.

- Grain size and distribution by microscopy
- Hardness by superficial hardness testers

Influence of temper and gauge Steel-based products are commonly used in gauges between 0.15 and 0.30 mm for cans and components and in tempers between 2 and 9, although thinner gauges, particularly DR material, are becoming more readily available.

Typical examples include:

	Temper	Gauge (mm)[a]
DWI 2-piece food can	T.2–T.4	~0.30
DRD 2-piece food can	DR8–DR9	0.16
Welded food can bodies	DR8	0.14–0.16
Food ends	T.5–T.7	0.20–0.21
Easy open end	T4/T.5	0.24

[a]Dependent upon container/component diameter.

5.3 Methods of container manufacture

The soldered three-piece tinplate can, standard up to about 1970, has been progressively replaced by

- Three-piece welded cans
- Two-piece cans made by either the drawn and ironed (DWI) route or by the draw-redraw approach (DRD)

One further three-piece method exists involving adhesive bonding of the side-seam, this method is dealt with briefly below.

This section describes the various can-making processes and reviews the factors which dictate the choice of the manufacturing route and the economics associated with the various options available. Brief mention is made of some key can-making equipment and processes used. The following is a brief description of the various can-making processes and serves to clarify the terminology used in the industry.

5.3.1 Three-piece can manufacture

Three-piece cans comprise a cylinder, today normally made by the welding route, and two ends. The essential steps in manufacture are as follows: cut up coil into rectangular sheets; apply protective lacquer (when appropriate), externally decorate (when appropriate); slit the sheets into rectangular blanks; form a cylinder and a side-seam; form flanges each end of the cylinder; fit makers ends, test, palletise and despatch to the filling lines. Such a can is commonly referred to as an open-top can (see Figure 5.4 which shows additionally the construction of the double seam closure of the ends). The formation of the welded cylinder is shown diagrammatically in Figure 5.5.

The side-seam as mentioned above is today normally resistance welded and this has been a development away from soldering. The change from soldering was on the basis of eliminating lead from the can contents and the factory

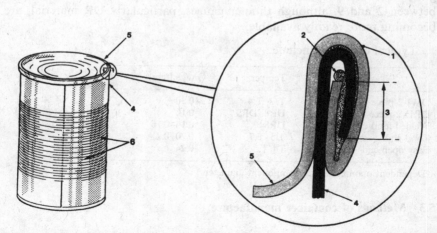

Figure 5.4 Open top (three-piece) food can and details of the double seam construction. (1) End curl; (2) lining compound; (3) seam lap; (4) can wall; (5) end counter-sink; (6) body beads.

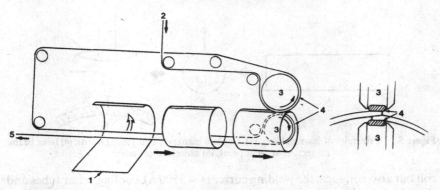

Figure 5.5 Stages in formation of a welded cylinder. (1) Blank rolled to cylinder shape; (2) copper welding wire loom; (3) welding rolls (electrodes); (4) copper wire contacts; (5) used wire to scrap or for recycling.

atmosphere since the most commonly used solder was 98% lead and 2% tin. An alternative solder was 100% tin which was normally avoided other than for baby food and soft drink cans because of cost. In addition to the technological considerations, welding uses less metal in the side-seam in that modern welders require an overlap of only approx 0.5 mm.

The minimum diameter than can in practice be resistance welded is limited to 52 mm. This is because the diameter of the cylinder needs to accommodate the welding current-carrying rolls, which are supported by the welding-arm (see Figure 5.6). The welding-arm not only mechanically supports the welding

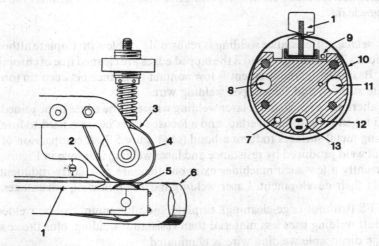

Figure 5.6 End of welding arm showing welding rolls and cross-section of the arm. (1) Z-bar; (2) side seam; (3) welding wire; (4) and (5) upper and lower welding rolls; (6) welded cylinder; (7) lacquer feed-pipe; (8) cooling water feed; (9) duct for nitrogen shroud (when required); (10) can wall; (11) cooling water return; (12) lacquer return pipe; (13) channel for welding wire.

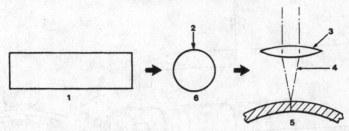

Figure 5.7   Principles of laser welding. (1) Tinplate blank; (2) butt joint; (3) lens; (4) laser beam;
(5) section at weld point; (6) unwelded cylinder.

roll but also conducts the welding current (~ 3500 Å), cooling water tubes and
the side-stripe lacquer supply in most instances. The need to be able to remove
the welded cylinder means the arm and the services it carries can be supported
only from one end.

Welded cans are made almost exclusively from tinplate although in
principle tin-free steel (TFS) can be welded provided the edges to be joined are
first cleaned of chromium oxide, which is necessary for two reasons. Firstly the
contact resistance between chromium oxide and the sacrificial copper wire
electrode used in the resistance welding process is too high and secondly the
abrasive wear of machine parts is excessive. Edge-cleaning of tin-free steel has
been attempted but is generally regarded as an unattractive operation in a
production environment.

Whilst tin-free steel cannot be conveniently resistance welded, it can be
readily welded by laser fusion. Thus the development of laser welding, of which
brief mention is made below, will permit the welding of TFS without the need
to edge-clean.

Laser welding   Resistance welding is really only viable with tinplate although
technically TFS can be welded if the lapped edges are cleaned free of chromium
oxide. Basically the requirement is low contact resistance between tin (on the
tinplate surface) and the copper welding wire.

An alternative approach is laser welding whereby the edges to be joined are
butted together, without overlap, and a focused laser beam is used to fuse the
touching metal surfaces to form a bond (see Figure 5.7). A comparison of the
types of weld produced by resistance and laser welding is shown in Figure 5.8.

Currently a few such machines exist but they are at a fairly rudimentary
stage in their development. Laser welding has the following advantages:

- TFS (without edge-cleaning), tinplate and aluminium can be welded
- Butt welding uses less material than resistance welding, plus the cost of
  the disposable welding wire is eliminated
- Laser welds are smooth, aesthetically more pleasing and are more easily
  protected
- Smaller diameters can be welded than is possible with resistance welding

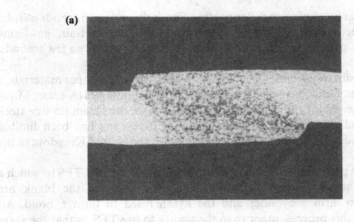

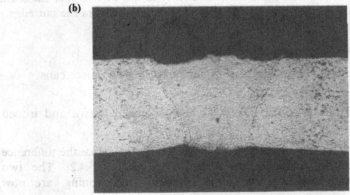

Figure 5.8 Comparison of the structure of (a) resistance (overlap × 100) and (b) laser (butt × 200) welds.

Currently, however, welding speeds are limited to about 35 m/min compared with 70 m/min by resistance welding. In addition the laser fused weld area is extremely hard resulting from the formation of 'martensite' during fusion and subsequent rapid cooling.

Irrespective of the method of welding, the side-seam normally requires protection of the exposed iron via a side-striping operation internally and sometimes externally. Further description of this process is given below.

Side-stripe application A number of methods have been devised for applying the side-stripe protection to the welded cylinder. These include the spraying of lacquers in both liquid (solvent-based) and powder form and the application of liquid lacquers by miniature roller coaters. Similarly various methods of side-stripe lacquer curing systems are employed including hot-air, gas-flame impingement and induction heating. Curing times are typically a few seconds.

Cement/adhesive side seams Driven by the ability to use cheaper materials, a number of companies have developed alternative technologies (A-seam, Mira-seam, Toyo-seam) for the manufacture of three-piece cans from tin-free steel. The use of adhesively bonded side seams for food cans has been limited principally to Japan, although some were made in the United Kingdom in the 1970s.

In brief, the process involves cylinders made from lacquered TFS to which a longitudinal strip of nylon is applied. The cut edges of the blank are overlapped to form a cylinder and the nylon fused to form a bond. An advantage of this process, other than the ability to use TFS, is that the seam does not require further protection as the fused nylon covers the cut edge.

5.3.2 Two-piece can manufacture

There are essentially two methods for manufacturing two-piece cans:

- Draw-redraw (DRD)
- Drawn and wall-ironed (DWI), sometimes called drawn and ironed (D&I)

although recent tooling developments are beginning to make the difference between these two methods less distinct (see Section 5.4.2). The two fundamental metal-forming operations, drawing and ironing, are now explained.

Drawing In the context of can-making, the process is one of cup formation, where the diameter is reduced at essentially constant metal thickness, by drawing metal from a flat sheet via a punch through a circular die (see Figure 5.9). In practice, some metal thickening occurs but in general terms the surface area of the cup is equal to the surface area of the blank from which the cup is made.

Redraw The diameter of the cup can be further reduced and its height increased by one or more redraw operations. The extent of reduction in diameter via redraw operation is governed by fairly rigid theoretical considerations [4] which are outside the scope of this book.

Figure 5.10 shows the draw-redraw operation to form a DRD can body, including the formation of a stronger base in the body-maker, and Figure 5.11 shows the stages in the manufacture of a DRD can.

The first stage in the manufacture of a DWI or DRD container is the formation of a cup, the basic tooling arrangement for which is shown in

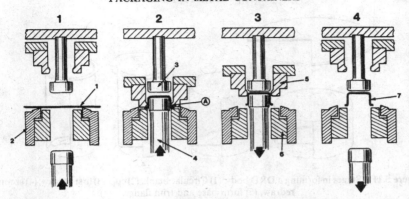

Figure 5.9 Typical tooling arrangement for a drawn can or cup. (1) Metal sheet or coil; (2) blanking tool (cut edge); (3) knock out pad; (4) ram/punch; (5) draw-die; (6) draw-pad; (7) drawn cup.

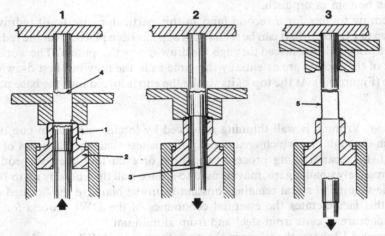

Figure 5.10 Typical redraw tooling arrangement. (1) Draw cup; (2) redraw sleeve; (3) punch; (4) base tooling; (5) redrawn container.

Figure 5.9. It will be apparent that a finished single-drawn container, with the introduction of base profile tooling, can be produced with the same basic tooling assembly.

Whilst Figure 5.9 shows the formation of a flange which needs to be subsequently trimmed, the tooling can be so designed to eliminate the flange as in the first stage of the DWI process. The latter approach is referred to as 'through-draw'. Stages 1 and 2 show the formation of the blank which becomes clamped between the draw-die and the draw-pad. The pressure holding the blank is carefully balanced against the force exerted by the ram to shape the cup whilst allowing the metal to flow controllably through point (A). (This area of the tooling can be designed to produce a cup without a flange.)

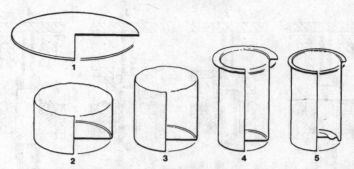

Figure 5.11 Stages in forming a DRD body. (1) Circular blank; (2) cup; (3) first redraw; (4) second redraw; (5) form base and trim flange.

Stages 3 and 4 show the fully formed cup with the draw-die separating with the cup positioned under the knock-out pad to allow ejection after withdrawal of the bottom ram/punch.

Redraw tooling for a second (and in this particular case final) redraw is shown in Figure 5.10. It can be seen that a cup has been placed over the redraw sleeve prior to being forced through the draw-die by the 'punch'. The working faces of the tooling are essentially the same as in the previous first draw/cup stage (Figure 5.9). At the top of its stroke the cup is forced into the base panel die.

Ironing Ironing is wall-thinning produced by forcing a redrawn cup on a punch through dies which create a gap that is less than the thickness of the metal. In a can-making process two, three or even four dies, to produce progressively smaller gaps, may be used to reduce wall thickness by up to 70%.

The volume of metal remains constant from the blank to the finished can and this fact dictates the essential economics of the DWI process for the manufacture of cans from steel and from aluminium.

Figure 5.12 shows the stages in the manufacture of a DWI can and it can be seen that the essential steps are the formation of cups from coil using vertical multi-toolpack presses. These cups are fed into wall-ironing machines where they are redrawn and ironed to form cylinders. It should be noted that the top of the can is allowed to thicken to facilitate flange formation and to assist stripping of the untrimmed can from the punch at the end of the ironing process.

After trimming the container to the correct height, it is washed and dried ready for the down-line operations listed in Table 5.5.

Figure 5.13 shows a typical toolpack comprising a single redraw die and three ironing dies.

The frictional considerations of drawing and ironing have important impact on the can-making process. The very high friction between metal and tooling and the extreme pressures produced in the ironing process necessitate flood

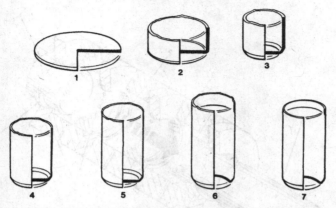

Figure 5.12 Stages in forming a drawn and ironed DWI body. (1) Circular blank; (2) cup; (3) redraw; (4) first ironing stage; (5) second ironing stage; (6) third ironing and dome forming; (7) trimmed body.

Table 5.5 Summary of steps in the manufacture of two-piece cans.

DWI	DRD
Unwind coil	Coil lacquer or cut to
Lubricate	sheets and lacquer
Blank disks and form cups	Blank disks and form cups
	Redraw (once or twice)
Iron walls and form base	Form base
Trim body to correct height	Trim flange or form flange
Wash and treat	
(passivate)	
Decorate (optional)	
External protection	
Internal protection	

lubrication and restriction of the process to tinplate and aluminium. In the former case the tin coating also behaves as a lubricant.

Attempts [5, 6] have been made to extend the wall-ironing process beyond tinplate and aluminium but these have failed to reach commercialisation.

5.3.3 Can ends

Can ends are of two basic types:

- Fixed (non-easy open)
- Easy open

In the case of food cans, fixed ends, i.e. those which need to be opened using a can-opener, are far more commonly used than easy open ends which are opened by removing a scored panel area (see Figure 5.14). This is in marked

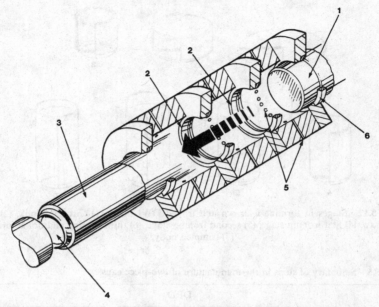

Figure 5.13 DWI toolpack. (1) Cup on punch; (2) ironing dies; (3) ironed/untrimmed can; (4) dome tooling; (5) coolant ports; (6) redraw die.

contrast to cans for beverages which up to 1990 at least have been closed with pouring aperture aluminium easy open ends (see Figure 5.15).

Fixed ends Ends for tinplate or TFS food cans are invariably made from steel, again either tinplate or TFS, and may be round, oval/irregular or rectangular to suit the shape of the container.

The end/body closure provides a hermetic seal via a double seam, the construction of which is shown in Figure 5.4. It will be seen that the seam construction involves a gasket material (lining compound) comprising essentially a solution of latex in hexane although water-based compounds are used for some applications.

The end is constructed to resist can internal pressure generated during retorting and importantly to recover its original profile on cooling. The latter property provides an indication of can spoilage, manifested as a bulged end, during storage.

The overall strength of the end is provided by the gauge and temper of the metal used in its construction and by the depth/shape of the countersink. Resistance to distortion by internal pressure is provided by one or more circumferential expansion rings (see Figures 5.14 and 5.16). Figure 5.16 also shows the tooling and initial formation of the double seam.

Easy open ends In order to provide convenience some cans are sold with easy open full aperture ends which are made normally from tinplate or, for some

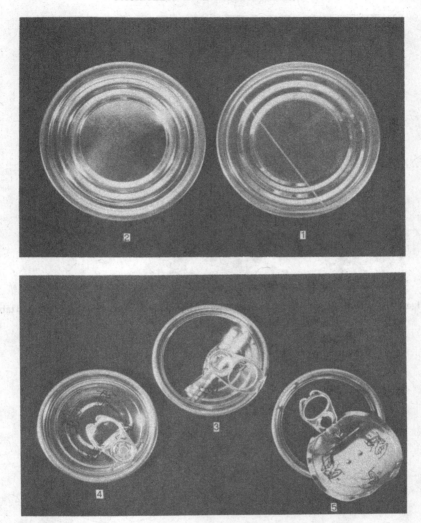

Figure 5.14 Fixed or non-easy ends (top) and full aperture easy open ends for food cans (bottom). (1) Lining compound; (2) expansion beads; (3) tab; (4) score area; (5) panel removed.

products, aluminium. The opening feature is produced by introducing a circumferential score close to the double seam and a tab which permits the panel contained within the score to be removed (see Figure 5.14).

Once the panel is removed the can contents can be poured (or otherwise removed) from the container. The removal of the panel leaves a torn surface on the panel and around the aperture which some designs seek to overcome by the introduction of safety-folds, either single or double (see Figure 5.17), with or without the introduction of guard beads.

The score can be placed internally or externally and normally requires

Figure 5.15   Two types of pouring aperture easy open ends for beverage cans. Left: Detached tab; right: ecology or stay-on tab.

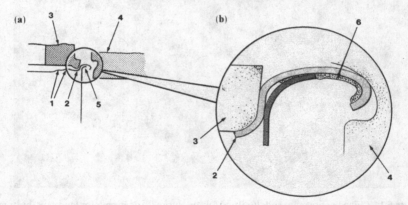

Figure 5.16   (a) Can flange and end within double seamer tooling; (b) detail showing formation of the double seam. (1) Expansion beads; (2) counter-sink; (3) chuck; (4) seaming roll; (5) flange; (6) lining compound.

repair either by spraying lacquer or oil or by electrophoretic lacquer application. The use of TFS is generally avoided because of unacceptable tool wear during the scoring and rivet-forming operation.

5.4   Selection of a can-making route

The following factors will influence the choice of the most suitable can-making route within a particular geographical area

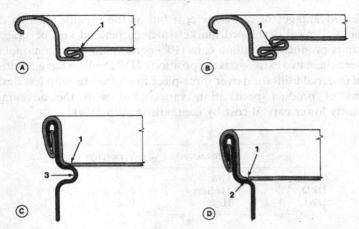

Figure 5.17 Single and double safety folds on full aperture easy open food ends. (A) Single safety fold; (B) double safety fold; (C) single safety fold with guard-bead; (D) single safety fold with necked-out body wall to provide guard. (1) Score; (2) necked-out guard; (3) internal guard bead.

- Products to be packed
- Food preservation method
- Numbers of cans to be made
- Variety of specifications
- Availability and price of raw materials

All the above need to be considered in any territory where significant investment in can-making is planned.

5.4.1 Product(s) to be packed

For food cans, where the contents are to be heat processed in the pack, both three-piece welded and two-piece (DRD or DWI) are relevant. If other than in-can heat processing is to be considered, e.g. aseptic packaging, then new rules begin to apply and a whole new range of containers and closures becomes available to the packer.

5.4.2 Size of the market and the manufacturing unit

The size of the market and of the manufacturing units required to serve that market will greatly influence the choice of the can-making route. Two important considerations are the capital cost and versatility of the equipment. Where the manufacturing unit is to be small and/or variety of can sizes and specifications large, welding is normally the preferred option. However, the need to provide lacquered sheets for some specifications must not be overlooked in considering the total manufacturing unit.

Where the manufacturing unit is large and variety is small, the DWI route will offer advantages in container cost but at the expense of high capital investment. Typical developed markets include petfood and beverage cans where units producing 1 billion cans (10^9) per annum are commonplace.

Between these two extremes is the position of DRD which has advantages in terms of material utilisation over three-piece manufacture, with less flexibility in terms of product/specification variety but with the advantage of significantly lower capital cost by comparison with DWI.

	Capital cost	Versatility
Welding	Low	High
DRD	Medium	Poor
DWI	High	Poor

These are broad generalisations and the choice of process may be further refined by e.g. the ability to purchase pre-coated coil, the availability of appropriate quality steel or aluminium and by the choice of tinplate or tin-free steel (TFS). These subjects are further developed below.

Which two-piece method? The strength of the can wall is clearly an important property governing filling, seaming of the closure, storage and resistance to abuse. This property is controlled principally by wall-thickness.

It will be obvious that if the same nominal can size is made by the DRD and DWI routes, then for the same nominal wall thickness, the ingoing gauge of metal will be substantially different. The example of the 0.5 kg can (typical diameter 73 mm, height 110 mm) illustrates the situation well.

Process	Ingoing gauge (mm)	Blank diameter (mm)	Wall (mm)	Base (mm)
DRD	0.18	179	0.18	0.18
DWI	0.30	154	0.13	0.30

Based upon the price structure of steel, which is related to gauge and area and not tonnage, the traditional rules suggest that

- High height/diameter ratio (tall cans) are favoured by DWI
- Low height/diameter ratio (squat cans) are favoured by DRD

In strict metal-forming terms and for like metals, these rules are generally applicable but other factors complicate the situation

- DWI necessitates the use of tinplate whereas DRD cans can be manufactured from prelacquered TFS (both types of container can be made from aluminium).

- DWI demands more costly equipment and hence greater volumes are required to justify the capital investment.
- The economics can be complicated further by partial wall-ironing of DRD cans.

Partial wall-ironing When taller cans are made by the DRD route it is possible to include an ironing die in the second redraw operation to reduce the wall thickness by up to approximately 20%. This requires sophisticated tooling but naturally improves the metal utilisation in the DRD process by allowing the use of a smaller blank ('cut-edge').

5.5 Mechanical properties of containers and ends

5.5.1 *General*

Food containers need to have sufficient strength and robustness to permit handling, filling and processing, storage and distribution without undue damage. The filled and closed containers also need adequate strength to withstand the temperature and pressure conditions used in the in-can sterilisation process.

The mechanical strength properties of (filled) containers are measured in terms of:

- Axial strength: resistance to buckling from top load
- Panelling resistance: resistance to external pressure/in-can vacuum
- Peaking resistance: resistance to in-can pressure

In practice, axial strength and panelling become closely inter-related in modern can body design. Peaking performance is accommodated via the design of the end(s) countersink depth and by the introduction of circumferential (expansion) beads.

5.5.2 *Axial strength*

If the three-piece can body, for example, is regarded as a perfect cylinder it can be assumed that axial strength is largely independent of metal gauge and temper within the range of loads that would be normally applied. In practice the can body does not have perfectly uniform and parallel sides and variations in symmetry and the presence of small imperfections have an effect on axial strength.

5.5.3 *Panelling resistance*

During the retorting process the external pressure generated in the retort is counter-balanced to a degree by the internal pressure generated within the can. The container specification is such that this differential pressure can be accommodated. In the extreme, the external pressure will cause the side walls of the container to collapse inwards. Cooling subsequent to in-can heat sterilisation of the can contents produces a vacuum within the can, which varies with products and filling conditions. This vacuum will exacerbate any tendency for collapse to occur.

Resistance to panelling is a function of gauge and strength of the side wall. Container diameters and height/diameter ratio will also have an effect. The tendency for panelling to occur can be substantially reduced by the introduction of circumferential beads into the side wall (see Figure 5.4). A great deal of fundamental work has been carried out to optimise the depth, number and spacing of such beads. However the most significant factor determining resistance to panelling is bead depth.

Introduction of such beads, whilst reducing the panelling by increasing hoop strength, significantly reduces axial strength. The physical properties of beaded and unbeaded 0.5 kg cans made from steel with a wall-thickness of 0.17 mm are as follows.

	Axial strength (lb)	Panelling (psi)
Unbeaded	1800	14
Beaded	900	26

In unbeaded containers increased gauge compensates for imperfections in the can symmetry and increases its resistance to mechanical abuse, this abuse subsequently reducing axial strength.

In beaded containers, the can designer is in greater control and can manipulate the various combinations of metal gauge, temper and beading configuration with greater certainty. This is of course of great importance as the manufacturer seeks to reduce container costs by lightweighting, i.e. the use of thinner and stronger metals. As a general guide-line beading is normally required for steel gauges below 0.20 mm and is essential for sound economics in two-piece food cans. The beads are introduced after the body maker, and in the case of two-piece cans, before internal lacquer protection is applied.

5.5.4 Peaking resistance

During the processing cycle, the contents of the filled container, including gas in the head-space and elsewhere, expand and generate an increased pressure within the can. Some release of the pressure can be accommodated by the concentric expansion beads designed into the end(s). Gauge and strength of the metal as well as detailed design of the end are important. There are, however, limitations to the use of very strong materials such as double-reduced tinplate and to some extent to the minimum gauges that can be utilised. These limitations are related to 'shape', rippling, warp or buckling, typically found in DRD materials and to producing adequate double-seams without splitting of the metal.

Distortion of the end during processing/sterilisation is acceptable provided it is not so great as to prevent the end resuming its original profile on cooling. Peaked ends are associated with can spoilage, either microbiological or from evolution of hydrogen gas resulting from chemical attack of the container by the contents.

5.5.5 *Measurement of mechanical properties*

A range of performance tests ensures that the mechanical strength of containers is always sufficient to withstand the loads imposed during the filling, seaming and processing operations, and subsequently during palletisation, storage and distribution.

The application of external overpressure in a sealed chamber measures the resistance to panelling during processing, while internal pressure is used to examine the expansion characteristics of can bases and ends, and determine the maximum pressure permissible prior to permanent failure.

Compression test equipment measures the maximum axial load-supporting capability of the cans, and pendulum impact tests are used to check resistance to, and effect of, damage.

At the design stage, these tests are related to measured properties of the container and material, and allow prediction of the behaviour of novel, or lightweighted products. Further performance checks during production ensure process control and product consistency.

5.5.6 *Secondary processes*

The chapter so far has concentrated upon the primary processes involved in the manufacture of the can body or body cylinder and only passing reference has been made to the secondary processes such as coating/lacquering, flange formation, necking, beading, etc. The latter have significant impact upon can-making costs and represent an area where significant development continues that will affect the economics of the future.

Beading of food cans Fundamentally the cost of a container and its axial strength are dictated by wall thickness. Panelling resistance can be improved substantially by introducing body beads into the side wall of the can and thus introducing increased 'hoop-strength'. Bead designs incorporate variations in the depth, contour, number and frequency. However the major factor influencing hoop-strength is bead depth.

Flange formation Flange formation at each end of three-piece cylinders and on the open-end of two-piece can bodies is essential for the formation of the universal double-seam closure (see Figures 5.4 and 5.16). The flanging operation can be by a die process or by spinning; the latter is becoming increasingly the standard for the industry.

Necking Originally all cans had straight walls whilst today diameter reduction to facilitate the use of smaller diameter and hence cheaper ends is commonplace for beverage cans. Parallel developments are taking place, notably in Europe, in food cans where the introduction of a stacking feature is seen as an additional benefit.

The necking operation can be carried out in a number of ways; by die-necking (single or multiple), by flow/roll-necking or most commonly by spinning in which case the flange can be produced at the same time. Spin-necking has the advantage over die-necking of using substantially lower forming loads and hence facilitating the use of thinner metal.

5.6 Coatings

A large proportion of food cans and ends have an internal and sometimes an external protective coating. In addition they may be externally decorated, in which case the external coating(s) may provide both a decorative and protective function.

5.6.1 *General classification*

Broadly speaking, surface coatings can be categorised as follows:

- Protective internal coatings: lacquers or sanitary enamels
- Pigmented external coatings: simple (white) coatings
- Clear external coatings: varnishes

These definitions are not absolutely water-tight and for example, external varnishes for draw-redraw cans are often referred to as external lacquers and indeed the basic chemical compositions are very similar.

These various coatings, whether protective or decorative, are generally applied as liquids comprising, in the simplest terms, a solution or dispersion of one or more resins/polymers in a solvent which may be organic or a mixture of water and organic co-solvent. In practice the formulation may be quite complex and involve a blend of resins, several solvents, plasticisers, catalysts to promote curing and additives to enhance flow and produce surface lubricity (waxes).

Such materials may be applied before or after fabrication of the container, dependent upon the method of manufacture, by roller coater or by spraying. More information about methods of application are given elsewhere.

Alternatively, surface coatings can be applied as powders, either thermoplastic or thermosetting, and subsequently fused onto the surface. As a means of reducing solvent effluents to the atmosphere, powders have advantages (being theoretically 100% solids) but their use has been limited by the high cost of manufacture and a rather high average particle size. Powder coatings are applied by spraying normally with an electrostatic assist.

For some very special end-uses, laminates, i.e. metal to which polymer films, such as polypropylene, have been adhesively bonded, are used. The use of this type of construction can be expected to increase in the future.

5.6.2 Protective internal coatings

Lacquer or (sanitary) enamel is the name given to those coatings which are normally applied internally to ensure compatibility between product and container. Broadly speaking lacquers can be regarded as being of two types, viz. oleoresinous and synthetic. The former are based upon natural products (fossil gums and drying oils) mixed with other resins. The latter are carefully synthesised products, which may additionally contain certain natural raw materials.

Oleoresinous products   These were amongst the earliest lacquers to be used in the canning industry and they are still in use, largely because of their low applied cost. Typical ingredients would include natural gums (rosin) and other resins with drying oils such as tung oil or linseed oil. In general terms they would not be regarded as high performance materials in that they lack process resistance and have poor colour characteristics. They are nevertheless suitable for a wide range of vegetable products such as green beans, butter beans when sulphur absorbing chemicals such as zinc oxide are incorporated in the formulation.

Synthetic products   Synthetic materials have replaced oleoresinous types and there is now available a wide range of lacquers designed to give specific performance with different products and for use in different can-making processes. There are few comprehensive texts relating to modern container coatings, largely because of the confidental nature of the formulations. However, the reader's attention is drawn to two texts dealing with the basic chemistry of resins [7] and modern coating types [8,9] and methods of application [10]. Some of the more common types and their uses are now described.

Epoxy lacquers   Epoxy resins, produced from the condensation reaction between epichlorohydrin and biphenol A (diphenylol propane), form the basis of a wide range of protective and decorative materials as well as adhesives (Scheme 5.1).

These materials are available in a range of viscosities and molecular

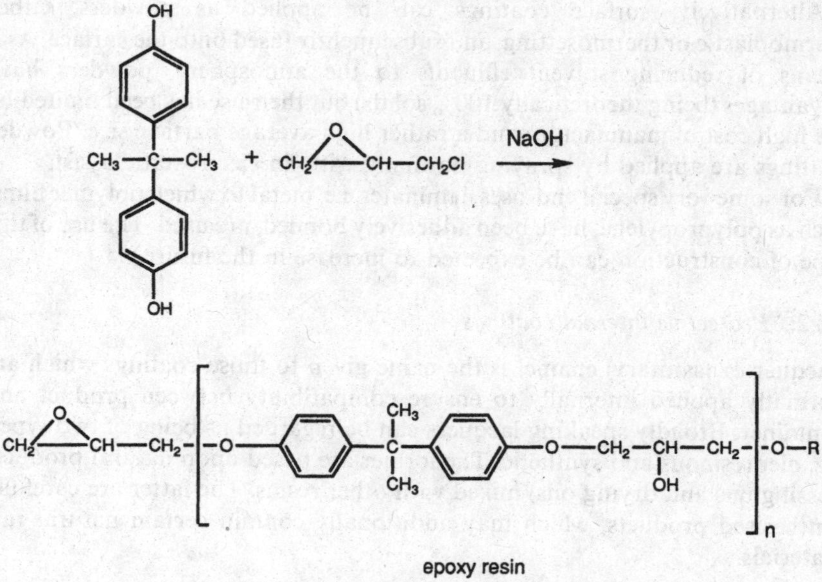

epoxy resin

Scheme 5.1

weights and are used in conjunction with other synthetic resin materials of which the following are common examples:

- phenolic
- ester
- polyamide
- amino (urea or melamine)

Epoxy-phenolic. Traditionally regarded as the mainstay of the food can industry, epoxy-phenolic lacquers combine high degrees of flexibility and adhesion with chemical resistance. The emphasis on flexibility or chemical resistance can be adjusted by the ratio of epoxy and phenolic content, the former giving flexibility and the latter chemical resistance.

Epoxy-phenolic lacquers, normally gold in appearance, are used for a wide range of acidic and non-acidic foods as well as for non-food products. They are sometimes used in a pigmented form (aluminium powder or zinc carbonate) to mask or absorb sulphide staining, they are also the base lacquer type for meat-release lacquers, in conjunction with waxes and aluminium pigmentation.

Epoxy-amino. Blending with amino resins, such as urea formaldehyde or melamine formaldehyde, produces coatings of high chemical resistance, which are nearly colourless (Scheme 5.2). Consequently they are used commonly for decorative purposes, where their lower stoving temperature is an additional advantage, and for internal protection of beverage cans.

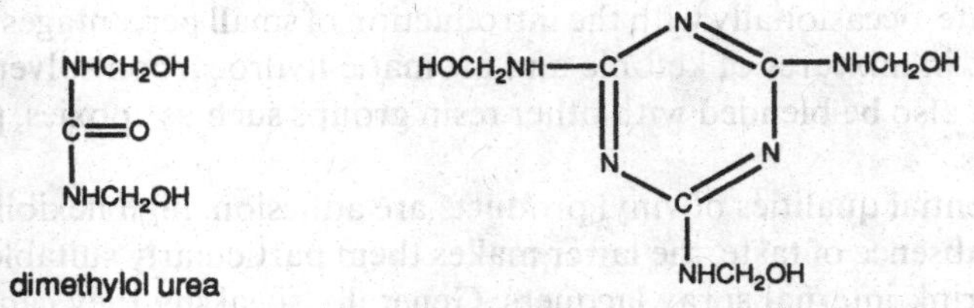

dimethylol urea

partially methylolated melamine

Scheme 5.2

Both resins are usually alkylated (commonly butylated) to varying degrees and cross-linking occurs with epoxy resin via methylol groups.

Epoxy ester. Epoxy resins esterified with certain fatty acids give rise to a family of lacquers and varnishes with excellent flexibility and colour. Their main use is in external/decorative varnishes for printed (food) cans.

Epoxy polyamide. When used in conjunction with polyamides, epoxies give rise to rapid curing systems normally requiring two-pack formulation, i.e. mixing of the ingredients immediately prior to use.

Polyamide resins are formed by the reaction between an organic acid and an amine; the generic reaction is as in Scheme 5.3. A well-known and typical polyamide is Nylon 6.6 produced from the reaction between adipic acid and hexamethylene diamine.

$$R\begin{array}{c}-COOH\\-COOH\end{array} + R^1\begin{array}{c}-NH_2\\-NH_2\end{array} \longrightarrow R\begin{array}{c}-COOH\\-CONH\cdot R^1-NH_2\end{array}$$

Scheme 5.3

In can coatings, however, polyamides with greater functionality/reactivity are used, e.g. to cross-link epoxy resins. In such cases polyamides formed for example from the condensation reaction between dilinoleic acid and diethylene triamine are used. The reactive amino group combines via the opening of the oxirane ring of the epoxy resin (Scheme 5.4).

$$CH_2\overset{O}{-}CH-CH_2 + H_2N-R \rightarrow -CH_2-CH-CH_2-NH-R$$
$$\underset{OH}{|}$$

Scheme 5.4

Vinyl lacquers. There exists a family of vinyl-based lacquers ranging from low viscosity/low solids solution vinyls, to mid-range vinyl alkyds used in decorative coatings up to the higher solids dispersion organosols.

Solution vinyl lacquers are solutions of co-polymer resin; vinyl chloride and vinyl acetate, occasionally with the introduction of small percentages of maleic anhydride, in mixtures of ketonic and aromatic hydrocarbon solvents. Vinyl resins may also be blended with other resin groups such as epoxies, phenolics and alkyds.

The essential qualities of vinyl products are adhesion, high flexibility and a complete absence of taste, the latter makes them particularly suitable for beer and soft drink internal spray lacquers. Generally speaking they can be dried (or cured) at very low temperatures and indeed on tinplate it is essential to keep below around 180°C to avoid catalytic decomposition by available iron. Common uses include deep drawn caps, decorative finishes and dry food packs.

Solvent resistance is generally very poor, particularly in the case of air-drying/unmodified vinyls, which actually re-dissolve in their own solvents. Resistance to steam sterilisation is also fairly limited.

Of increasing use in the manufacture of drawn food cans and ends are organosols. Organosols are dispersions of high molecular weight polyvinyl chloride (PVC) resins in hydrocarbon solvents with the inclusion of a suitable plasticiser, e.g. dioctyl phthalate and resin additives. Commonly polyesters and acrylic resins are used as adhesive promoters to metal, and phenolic or amino resins are used to introduce some degree of cross-linking and hence chemical resistance. Being dispersions they can be produced at much higher solids (50–70%) at relatively low viscosities than is possible with solution vinyls (Scheme 5.5).

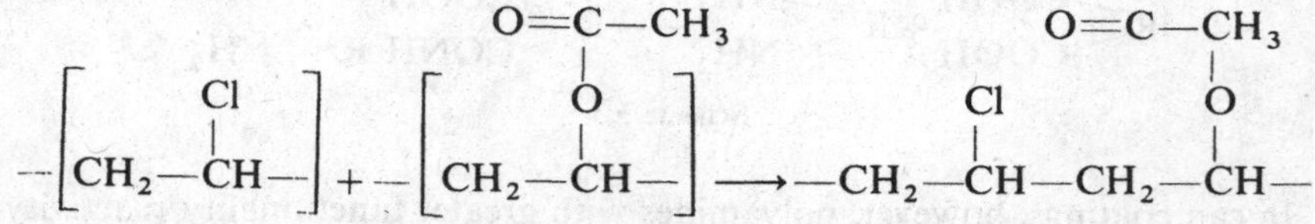

$$\left[\begin{array}{c} Cl \\ | \\ -CH_2-CH- \end{array}\right] + \left[\begin{array}{c} O=C-CH_3 \\ | \\ O \\ | \\ -CH_2-CH- \end{array}\right] \longrightarrow \begin{array}{c} Cl \quad\quad O=C-CH_3 \\ | \quad\quad\quad | \\ -CH_2-CH-CH_2-CH- \\ \quad\quad\quad\quad | \\ \quad\quad\quad\quad O \end{array}$$

Polyvinyl chloride + polyvinyl acetate ⟶ Vinyl chloride–vinyl acetate co-polymer

Scheme 5.5

Organosols have all the desirable vinyl properties of absence of flavour, flexibility and adhesion but in addition they have better process resistance and hence can be more widely utilised. In the context of process food cans they have reasonable chemical resistance and resistance to sulphide staining but they are prone to absorb food colourants. They can be applied at low viscosity but with high solids either unpigmented or pigmented with aluminium.

Their drying performance is quite different from other lacquers and is a three-stage process. The first stage involves solvent evaporation, the second involves fusion of the PVC particles into a coherent film and finally a cross-linking reaction occurs.

The third common category of vinyl lacquers/coatings are the vinyl alkyds (see below), which are used principally for deep drawn white coatings and as a size-coat to promote adhesion between other coatings and the metal substrate.

The drying of these materials reflects their chemical composition in that they dry partly by solvent evaporation and partly by oxidation/heat polymerisation.

Phenolic lacquers. Phenolic lacquers were one of the earliest types of material to be synthesised by the reaction of phenol with formaldehyde (Scheme 5.6). The reaction under alkaline conditions with excess formaldehyde results in ortho- and para-substitution into the aromatic ring of the phenolic. Subsequent heating allows polymerisation to occur via a series of methylene bridges and via ether links. Such materials found wide usage because of their extremely good chemical resistance and resistance to sulphide staining. However their use requires very careful control of the thin films applied and of the stoving temperatures 190–195°C. Under stoving limits chemical resistance whilst excessive stoving increases the brittleness of the film. Poor flexibility characteristic of phenolic lacquers limits their use to three-piece bodies and ends.

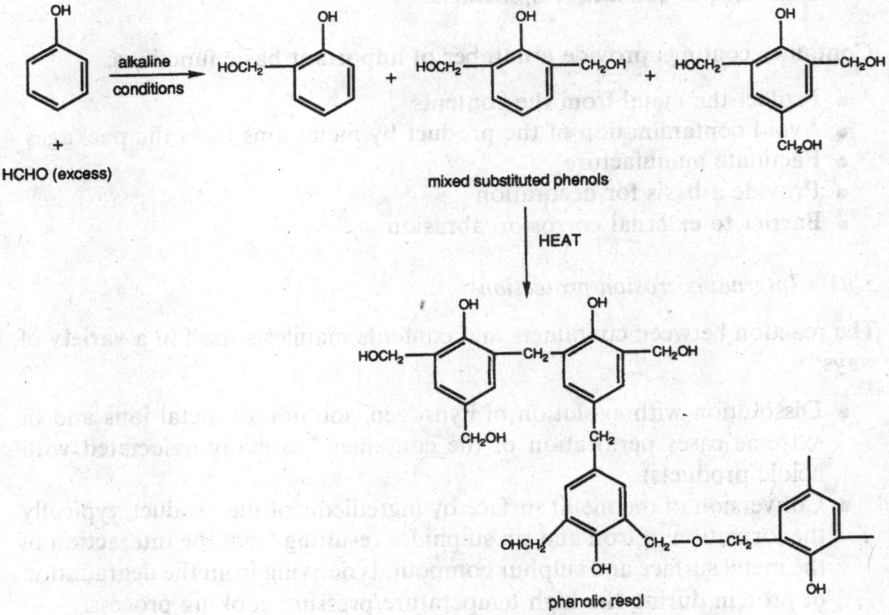

Scheme 5.6

Acrylic lacquers and coatings. The most common usage of acrylics is the manufacture of pigmented and clear decorative coatings where high temperature resistance and/or resistance to steam sterilisation is required. A typical example would be decorated solid meat packs where high stoving of the meat-release lacquer (applied last) to effect maximum lubricity is required.

More recently internal white lacquers have been formulated as alternatives to white vinyls, which have relatively poor heat resistance.

Alkyds and polyesters. Alkyds formed by the esterification reaction between glycerols, such as glycerol and pentaerithrytol, and phthalic anhydride are suitable only for external decorative coatings and inks because of the taste characteristics introduced by the oil content. Alkyds are readily modifiable with other resins such as vinyls to give a range of products with very wide properties in terms of adhesion, gloss and flexibility.

Closely related and finding increasing usage are other types of polyesters which are oil-free and based upon, e.g. isophthalic acid; they are used singly or in combination with other resin types such as phenolics. Pigmented polyesters with good colour retention on stoving are used for drawable white lacquers and varnishes, whilst in combination with phenolic materials they are sold as cheaper alternatives to epoxy-phenolic lacquers.

5.7 Functions of can lacquers/enamels

Container coatings provide a number of important basic functions:

- Protect the metal from the contents
- Avoid contamination of the product by metal ions from the packages
- Facilitate manufacture
- Provide a basis for decoration
- Barrier to external corrosion/abrasion

5.7.1 *Internal corrosion protection*

The reaction between containers and contents manifests itself in a variety of ways

- Dissolution with evolution of hydrogen, solution of metal ions and in extreme cases perforation of the container (normally associated with acidic products)
- Conversion of the metal surface by ingredients of the product, typically the formation of iron and tin sulphides resulting from the interaction of the metal surface and sulphur compounds deriving from the degradation of protein during the high temperature/pressure cooking process.

The above broad generalisations can be further subdivided to describe the types of interaction between product and container, as follows:

- Tin sulphide staining
- Iron sulphide staining
- Selective and severe tin dissolution
- Acid attack leading to hydrogen evolution, high metal content and ultimately perforation

- Staining by natural and artificial colourants
- Beneficial tin dissolution

The corrosion mechanisms of the above are dealt with elsewhere and here we will concern ourselves only with product and lacquer selection aspects with respect to different containers and components.

Always there is a need to strike the correct compromise of price, performance, container specification (beaded/unbeaded) and product. In the case of tinplate cans, the tincoating level is also an important consideration. Table 5.6 summarises some typical product/lacquer combinations for three-piece welded cans.

Table 5.6 Some examples[a] of three-piece welded tinplate can body specifications for a variety of food products.

Lacquer specification	Tincoating	Products
None (plain tinplate)	High (~ 11.2 g)	White fruits: grapefruit, pears, peaches and pineapple Vegetables: asparagus, tomatoes, artichokes
Single-coat epoxy-phenolic or phenolic	Low (~ 2.8 g)	Meats: chicken, fish, duck Vegetables: green beans, peas, spinach
Two coats: oleoresins plus phenolic	High (11.2 g)	Gherkins, concentrated grapefruit juice, strawberries, red plums, damsons, beetroot in vinegar
Epoxy with meat-release additive (wax)	Low–medium (5.6 g)	Solid meat packs: ham, tongue, luncheon meat

[a]These are examples of general specifications in commercial use and serve to demonstrate differences in pack requirements. Other specifications are used which may vary in detail from the examples chosen here.

5.7.2 Protection of the product

Internal corrosion and product contamination are in many cases complementary processes. In the latter case, product contamination does not always result in a deterioration of the nutritional qualities but frequently the organoleptic qualities are affected.

Thus, for example, dissolution of iron by beer or soft drinks affects the flavour at quite low levels and consequently packers specify very low levels of iron and aluminium in beverage cans. With few notable exceptions (e.g. lemonade), tin pick-up is not a problem in carbonated drinks.

Generally speaking food products, as opposed to beverages, are more tolerant of metal pick-up levels and only in the case of tin is there a recommended guideline level of 250 ppm. Some food packers set levels of their own, e.g. for iron, although in practice for acidic products in particular, high levels of iron dissolution could inevitably result in unacceptable levels of hydrogen evolution and 'hydrogen swells'.

Other metals of concern are lead (note: British Standard BS3252 (1960) dictates < 0.08 w/w lead in electrolytic tin coatings), where in most cases lead is limited to 0.2 ppm, chromium and copper. In all cases the can maker/canner would be looking to achieve levels between 'none-detectable' and subparts per million [11].

The reader is directed to alternative literature for further specialist information on the subject of metals in food and beverage packaging.

Examples of requested levels (i.e. customer expectations) for metals in carbonated drinks currently are for iron 0.3 ppm and 1 ppm after 6 months storage for beer and soft drinks, respectively, and for aluminium 2 ppm.

Some special and extreme interactions between tin and product exist. In the case of white fruits, relatively high levels of tin dissolution are required (i.e. 'free tin') to maintain colour and taste. Other products such as asparagus, beans and pasta products in tomato sauce also have this requirement although this is very much a matter of taste. Products packed in plain cans in the United Kingdom are packed in lacquered cans in other countries.

5.7.3 *Facilitating manufacture*

Most metal-forming processes require some form of lubrication. In the case of wall-ironing, it is provided in the 'bodymaker' by oil/water emulsions and in the case of tinplate containers, the surface tin facilitates the ironing process.

Other metals such as tin-free steel cannot be worked economically either with or without lubricant because the surface of such material is so hard and abrasive. The manufacture of draw-redraw cans and food can ends from TFS necessitates the use of prelacquered steel or coil. Very flexible vinyl materials are commonly used for DRD can manufacture and these lacquers incorporate internal lubricants (e.g. waxes).

Drawn containers   In the simplest terms, DRD cans cannot be made from uncoated TFS and such containers are normally not an economic proposition when made from tinplate (or aluminium?). Such containers (DRD/TFS) are made from either precoated coil or precoated sheet in order to protect the expensive tooling and can handling system from the abrasive wear by the chromium oxide layer on the plate.

Two other properties are required by the coated plate:

- Flexibility/lubrication
- Product compatibility

High flexibility normally is associated with loosely cross-linked/high molecular weight materials such as vinyl organosols and certain polyesters. However, these are not necessarily the first choice for chemical resistance where the selection would be epoxy-phenolic and phenolic lacquers. By and large the surface coating manufacturers have responded with good compromises and this, in combination with high film weights, provides one solution for most products. Some products cannot, however, be satisfactorily packed in as-made containers and a secondary 'repair' spray lacquer is applied. Such products would include highly acidic packs.

Lubrication is also an important factor. In addition to flexibility, the lacquer is normally internally lubricated by the addition of waxes, such as polyethylene and paraffin types, which migrate to the surface when the lacquer is stoved.

The final consideration is method of application. Two methods are possible, coil coating and sheet-fed coating. The former offers the advantage of facilitating coil feeding of the presses used to make cans as well as producing a smoother film, by reverse roller coating, which is beneficial to the metal-forming operation. Coil-coated steel can of course be cut to sheet for sheet-fed press operations.

The inference of this is that viscosity control is extremely important to ensure good smooth application of the plate. The availability of coil-coated plate is a matter of concern since, whilst fairly widely available in the United States, there is relatively little capacity for can coatings in Europe and other territories. This is expected to change in the future.

Ends The arguments concerning tooling design also apply to the manufacture of ends from tin-free steel. However, the metal-forming operations are less severe and this permits the use of a wider range of coatings; this range would include organosols applied at lower film weights than for DRD.

There tends to be a wider range of specifications for ends than for DRD cans and today the bulk of ends are manufactured from sheet-coated TFS or tinplate. Broadly, however, the following rule applied. For acid products, organosols are preferred at the expense of some lack of resistance to natural colour staining, for example from tomatoes and other 'red' fruits. For non-acid products lower film weights of epoxy-phenolic or polyester phenolic or even oleoresinous lacquers are used. In the future we can expect to see lacquer rationalisation perhaps accompanied by the availability of coil-coated TFS.

5.7.4 *Base for decoration*

Coatings, usually pigmented with titanium oxide or other pigments and applied in thick films > 10 μm, are used to provide the background for printed decoration. These materials frequently need to be formulated to allow subsequent forming operations.

5.7.5   *External corrosion and abrasion resistance*

Both tinplate and other steel-based products visibly rust. Aluminium containers discolour and are highly prone to acid attack as in the case of secondary corrosion in the soft-drink can, notably in hot climates. External protective coatings (varnishes) are applied as a means of providing abrasion resistance and as a barrier to external corrosion.

The above categories are not sharply defined and frequently a material may be required to perform more than one function. Unless very considerable cost penalties are to be borne, the formulation of a coating performing more than one function is invariably a compromise. Typical examples include the highly flexible vinyl organosols used for deep (DRD) containers which have limited stain resistance to, e.g. sulphide or natural colours. A second example is the formulation of a whole family of epoxy-phenolic lacquers where the percentage of phenolic ingredient (which has high chemical resistance but limited flexibility) is varied to suit the needs of flexibility from the epoxy ingredient and chemical resistance of the phenolic.

5.8   Methods of lacquer application

Three basic coating processes are in common use:

- Sheet coating by roller coater
- Coil coating by reverse roller coater
- Spraying

Further information on the various processes can be obtained elsewhere [9, 10] but particular advantages and uses of each method are outlined below.

5.8.1   *Roller coating in sheet form*

This process is essential for sheet coating of tinplate for welding cans since the process allows for stencilling, i.e. the provision of plain margins (see Figure 5.18) which permits resistance welding of the side seam. Plate for sheet-fed DRD lines is also prepared in this way.

5.8.2   *Coil coating*

This process can only be justified for applications where very large volumes of material of essentially the same specification are required. Capital cost is high, as is the cost of stoppages to effect either substrate or coating changes.

Typical applications are TFS for DRD cans in the United States although its use for food can ends is anticipated. In addition to cost considerations, lacquering in the coil permits the use of reverse roller coating which produces a more uniform film ideal for drawn can manufacture. Stencilling necessary for welding cannot be achieved in the coil process.

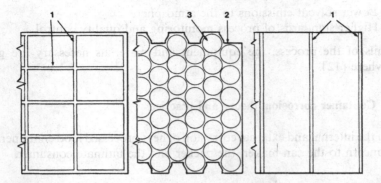

Figure 5.18 Various sheet lacquering lay-outs (stencil lacquering). (1) Unlacquered margins/areas; (2) scrolling to optimise metal utilisation; (3) spot lacquer for ends or DRD disks.

5.8.3 *Spraying*

Spraying is used for made-up containers (DWI) and as one means of applying side-stripe protection on welded cans. Wider descriptions of the processes are available in other texts [6, 7] where printing and decorating are also covered. Mention in this chapter is made primarily to indicate the type of processes used and how their availability and the materials to support them can affect the choice of can-making route.

Broadly speaking the following rules apply:

- Welding: requires availability of, or access to, sheet-coated (stencilled) tinplate sheets
- DWI: requires the skills and materials necessary to operate spraying machines
- DRD: requires either sheet roller-coated plate (all-over lacquer for TFS or spot lacquer for tinplate), or
- coil-coated coils or sheets from coil-coating operations

In a green-field site, the need to purchase sheet lacquering equipment as well as can-making equipment may influence the choice between DRD and DWI.

5.8.4 *Electrocoating*

Electrophoretic deposition of protective coating films has been well established in the automotive industry for many years. More recently it has been developed as an alternative for applying protective films to finished can bodies, such as DWI and DRD cans made in tinplate or aluminium. No machines are known to be in commercial production but the position can be expected to change. The advantages of the process are seen to be

- Lower materials usage; thinner, more uniform lacquer films

- Lower solvent emissions to the atmosphere
- Higher standards of process monitoring and quality control

Details of the process, the equipment and materials necessary are given elsewhere [12].

5.9   Container corrosion: theory and practice

Both the internal and external corrosion of heat processed food containers are of concern to the can-maker, the packer and the ultimate consumer.

5.9.1   *External corrosion*

In the case of tinplate, which is the dominant material used for heat processed containers, external corrosion results in either the formation of rust or localised staining and detinning associated with contact with aggressive alkaline compounds. These effects can be minimised by good control in the packer's plant and in distribution, e.g. over-cooling of cans after processing can initiate external rusting. Excessive use of chlorine in cooling water can also cause the corrosion of tinplate. These together with the prevention of boiler-water 'carry-over' should be avoided in cannery practice. The over-wrapping and storage of processed cans are also areas where good canning practice needs to be implemented if external corrosion is to be prevented.

Bright or labelled cans should never be stored 'wet' or warehoused where temperature and humidity can fluctuate to such an extent as to allow the cans to 'sweat'. In areas of high humidity, this phenomenon can occur when temperature differences of 7°C occur between the temperature of the cans and the surrounding atmosphere. Where shrink-wrapping is employed, external rusting is minimised when cans have been packed dry.

Aluminium is, by and large, less problematical in terms of external corrosion in that its use is restricted mainly to shallow-drawn containers for food and paté. External corrosion in the case of processed food containers manifests itself principally as staining.

5.9.2   *Internal corrosion*

Both steel and aluminium present specific difficulties in terms of internal corrosion. Thus, with the notable exceptions of products requiring 'available tin' for organoleptic reasons, most cans are internally lacquered to limit the interaction between container and contents. For products requiring tin, only the welded tinplate can is an option as, by and large, unlacquered DWI tinplate cans do not give satisfactory pack performance.

Internal corrosion and the mechanism by which it proceeds varies with the nature of the product and the type of can. Dissolution of the component metals of the container results as corrosion progresses. The consequence is that the container may either swell or perforate resulting in commercial spoilage. In addition to this type of corrosion, other forms of attack occur resulting in either staining of the container or the product or changes in the colour and flavour of the product. The most common in this category is the formation of iron and tin sulphide.

When reviewing internal corrosion there are two significant aspects to be considered. The first is the effect of internal corrosion on the commercial acceptance and marketability of products and secondly, and more importantly, is its effect upon public health.

We consider first the case of plain tinplate containers which have been traditionally used for a number of acid and non-acid products. In these cases, the internal corrosion which has occurred has been considered of benefit to the palatability and visual acceptance of the product. When these products are packed into plain cans the residual oxygen in the can, derived from the atmosphere or the product, either trapped in its tissues or absorbed on its surface, or both, acts as a de-polariser. Dissolution of tin occurs without hydrogen formation. The oxygen is rapidly used up in the can when the next stage in the corrosion process occurs. Here other inorganic or organic de-polarisers become involved when the tin continues to dissolve. Hydrogen is usually evolved only when considerable detinning has occurred exposing appreciable amounts of steel.

The reaction is then an electrochemical one with the tin affording anodic protection to the steel (see Section 5.9.3). The rate at which the reaction occurs is dependent upon the composition of the steel, the residual oxygen content of the can when closed, the presence of other depolarisers or corrosion accelerators, e.g. anthocyanin pigments and nitrates, and the temperature of storage. The storage temperature has a considerable effect on corrosion and generally it is recognised that the rate of corrosion almost doubles with an increase of 10°C.

While the acid content of a canned food has a definite effect on its corrosivity, its effect is seldom directly proportional to its concentration as measured by titration or pH value. A product's tendency to promote or inhibit corrosion depends more on the nature of the types of acid present.

Certain trace metals can also influence the type and rate of corrosion. 'Excess' copper is an example; its presence may cause swells in plain cans and perforation in lacquered cans. When present in sufficient amounts in acid products, the corrosion is normally accompanied by the 'plating out' of the copper onto the metal surface of the container. Corrosion involving plating out normally occurs when the copper levels are at about 4–5 ppm. However, in products such as canned apple juice and soft drinks, corrosion has occurred where levels as low as 0.5–2 ppm of copper have been present.

Nitrates, which are natural constituents of many foods, have been known to greatly accelerate corrosion in acid products packed in tinplate containers.

Below pH 5.4 nitrates act as de-polarisers. However under certain conditions the corrosive nature of nitrates is influenced by the presence of other de-polarising substances such that their corrosivity can be greatly increased or decreased. As a consequence it is very hard to predict the corrosive action that may result merely by observing the nitrate content of the product. Equally, because the nitrates are broken down into other products during corrosion, the amount of nitrates in a product, after corrosion has occurred, will not indicate the original nitrate content of the product. However, it is good manufacturing practice to get as low a figure as possible and especially in the case of acid products to pack at initial levels of less than 8 mg/l measured as nitrate-nitrogen.

There are also toxicological consequences of nitrates in water supplies. Methaemoglobinaemia, the conversion of blood haemoglobin to meth-aemoglobin, results in the loss of the transportation of oxygen to the body tissues. The conversion process can be affected by the presence of nitrate, which may in turn derive from nitrate arising in food or water. Infants are particularly susceptible to methaemoglobinaemia which could result in cyanosis and death. For this reason the World Health Organisation has recommended a limit of 45 mg/l measured as nitrate ion in water supplies (equivalent to 11 mg/l as nitrate nitrogen).

Although the result of corrosion detracts from the internal appearance of the plain container, the corrosion has in part been considered beneficial to the flavour and colour of certain products. Dissolution of tin in the container produces a reducing effect which protects against oxidative taints and also imparts the characteristic 'bite' to products such as citrus juices, white fruits and subacid products, e.g. beans in tomato sauce. It is considered that only when the tin concentration reaches several hundred parts per million can metallic tin actually be tasted. The presence of dissolved iron in the product is indicated at smaller quantities, e.g. 50–60 ppm, indicating that corrosion is far advanced.

The colour of the product is also favoured by the reducing action of the tin and in addition to the products already mentioned, some vegetable packs, e.g. mushrooms, asparagus, potatoes and carrots have also been traditionally packed in plain containers.

In those products which contain anthocyanin pigments, however (e.g. 'coloured' fruits), the reducing action of the tin has an unfavourable effect and the colour changes produced by the reduction of the pigments would result in a commercially unacceptable product. This effect is overcome by the use of lacquered containers for this range of products.

Aesthetically undesirable corrosion effects occur frequently, resulting from reactions with tin and iron; these are classified under the heading of sulphiding. Thermal degradation of certain organic sulphur compounds in foods can result in the formation of stannous and ferrous sulphides where these metals are exposed. In the case of stannous sulphide, the result is a fixed blue-black discoloration of the tinplate. In the case of ferrous sulphide the result is the formation of a black loose deposit usually associated with the headspace and in those products where the pH is above 6.0. To protect tinplate against sulphiding, special lacquers which afford good coverage and sulphur absorbent properties, such as the inclusion of zinc oxide or zinc carbonate, are used.

In the case of acid products, the presence of sulphur dioxide and sulphur can also present serious corrosion problems. They are reduced to hydrogen sulphide at tin and iron, producing severe odour and flavour problems in canned soft drinks and wines. By acting as de-polarisers they can also enhance the rate of metal dissolution and sulphur dioxide has been the cause of rapid detinning problems in plain cans. Very small quantities, as little as 2 mg/l, have been known to greatly reduce the shelf-life of tinplate containers.

Where lacquered containers are used, small areas of tin and iron may be exposed at pores or scratches in the lacquer. The beneficial action of sacrificial detinning is rapidly lost and the iron is soon able to corrode freely. Failure of this container for acid products is generally associated with the time taken for the onset of hydrogen swell formation, the continuing corrosion process then resulting in can body perforation. The commercial significance of this therefore is to produce containers with the minimum of metal exposure by providing good lacquer coverage. As previously mentioned, those products which are discoloured by contact with tin have of necessity to be packed into fully lacquered containers.

It is important to note that whilst aesthetically undesirable, the effects of corrosion resulting in either colour or flavour changes in the product or producing a visual defect in the container, do not significantly affect public health.

Where container corrosion is implicated in its effect on public health, this is more often than not associated with the dissolution of trace metals, and tin and lead in particular. At present, the recommended limit for tin in foodstuffs is 250 ppm. For most canned foods the lead limit is 2 ppm (reducing down to 1.0 ppm) but for baby-foods it is 0.5 ppm and soft drinks 0.2 ppm (in the United Kingdom only). In practice, the move from soldering to welding ensured that levels of lead above 0.2 ppm will not result from the interaction between container and product.

In the case of tin, it is internally plain containers which obviously come under scrutiny where tin dissolution progresses with storage time. The rate of pick-up is relatively high initially, with a gradual tailing-off on storage. Considering the volume of cans produced there is little evidence of illness arising directly from the ingestion of canned foodstuffs containing high concentrations of dissolved tin. Very high levels are needed to be present in the product before a definite metallic taste is detected. It is likely that the product would be rejected prior to consumption from either the colour or odour of the product or the visual appearance of the container. However, it is believed that where illness has been caused it has resulted in vomiting probably produced by irritation of the mucous membranes of the gut by the tin. High levels of tin have been associated with high nitrate levels in the water used in the production of the canned products.

Should legislation change with regard to the reduction of the limit of tin concentrations permissible in canned foodstuffs then there are two possibilities which might arise. The first would be a drastic reduction in the shelf-life of many products at present packed in plain containers. The second, and probably the more likely result, would be the increased use of fully lacquered containers. This would result in slight but probably not commercially unacceptable changes in the colour and flavour of some products and would of course alter the basic corrosion mechanism in the container. The use of tin bearing lacquer systems to provide controlled amounts of tin going into solution, in an attempt to provide the desirable colour and flavour changes in the product, has been resisted.

When soldered cans were the norm, the use of fully lacquered containers played an important role in limiting the amount of lead pick-up in the container. Whilst soldered food cans are still manufactured, their number is reducing and by the year 2000 all are likely to be welded or of a two-piece construction. Thus lead pick-up is not discussed further.

5.9.3 Theory

When considering the corrosive interaction between a metal container and its contents, it is important to realise that we are considering electrochemical reactions involving electrodes (container and lid/closure) and electrolytes (products). The precise reaction that occurs will be influenced by numerous factors, including the number and type of metals present, the type of product and the presence or absence of air within the pack. There are, however, some general rules which apply universally.

Rules governing corrosion in metal cans   (1) Anodes are areas of the can which dissolve or oxidise, i.e. electrons are lost. Similarly cathodes are the areas where reduction reactions occur, i.e. electrons are gained. Thus the reaction will involve the flow of ions in the product and electrons in the metal of the container. At any moment in time, the anodic current is equal to the cathodic current, the significance of this statement is further explained in the context of 'pitting' corrosion.

The reader must also recognise that the surface of the metal container, e.g. one made from tinplate, is not a simple one and on micro-scale it will contain areas of tin and of exposed iron and other metals. Whilst the anodic dissolution of metal, for example, is obvious, the cathodic counter-reaction is often less obvious.

(2) The reactions occurring at anodes and cathodes will in effect also obey Faraday's Law. The weights of various metals which dissolve are related to

current which flows and the electrochemical equivalent of the metal incurred. Thus if we consider a detinning reaction (tin being in this situation anodic), then if the amount of tin dissolved is known, the total current flow (coulombs) can be determined as can the amount of oxygen reduced at the cathode.

$$Sn \longrightarrow Sn^{2+} + 2e^-, \quad \text{anodic reaction}$$

$$\left.\begin{array}{r} 2H^+ + 2e^- \longrightarrow H_2 \\ 2H^+ + \tfrac{1}{2}O_2 + 2e^- \longrightarrow H_2O \end{array}\right\} \quad \text{cathodic reactions}$$

Pitting corrosion 'Pitting' or localised corrosion, which in the extreme may lead to perforation, is localised electrochemical attack where the balanced electrode currents are related to very unbalanced areas of anode and cathode. A good example would be an unlacquered tinplate container with a lacquered aluminium end with a product pack which renders the aluminium end anodic. We have a situation where the unlacquered tinplate surface of the can body is large and cathodic and the imperfections in lacquering of the aluminium end render very small surface areas of the end anodic. Since the anodic and cathodic currents are balanced and equal, it follows that the cathodic current density on the tinplate is very low and the anodic current density on the exposed aluminium very high. Hence solution of the aluminium will occur locally at a very high rate leading to 'pitting' and ultimately, perforation.

Cathodic reactions Whilst the results of anodic reactions are fairly obvious, e.g. metal dissolution with hydrogen evolution or sulphide staining, those of cathodic reactions are less obvious and sometimes ignored. However, a thorough understanding of these cathodic reactions will lead to an increasing understanding of the factors controlling shelf-life and minimising can corrosion. Anodic corrosion can be cathodically controlled.

Common cathodic reactions include the reduction of oxygen and nitrate, reduction of sulphurous compounds (sulphur dioxide, sulphur and sulphur-containing proteins in e.g. meat, fish and some (leguminous) vegetables). The latter group gives rise to the characteristic black sulphide staining with tinplate containers.

Limitation of the corrosive anodic reactions and hence long shelf-life are promoted by packing conditions that limit the availability of cathodic reactions.

Choice of container materials and their effect on corrosion Tinplate remains the dominant material from which food cans and ends are made although the use of tin-free steel for fixed (non-easy open ends) and draw-redraw bodies is increasing. A second trend is that the levels of tincoating are reducing, to be compensated for by progressively more effective protective lacquer systems. Thus whilst tin weights exceeding 10 gsm were once commonplace, today tin weights as low as 1–2.5 gsm are used, the limit being frequently dictated by the ease of resistance welding rather than electrochemical performance. The use of higher tincoatings is retained for unlacquered containers where tin dissolution is required for organoleptic reasons or for very acidic products such as red plums, beetroot and gherkins in vinegar and concentrated grapefruit juice.

Tinplate containers   In general terms, acidic products are more corrosive to tinplate than non-acidic ones although no strict relationship exists between pH and rate of corrosion. Acidic detinning (tin dissolution) has been studied in depth and whilst the number of publications is great and their review outside the scope of this publication, it can be concluded that the basic rules of corrosion (described above) apply. Since acidic fruits represent a high proportion of the world's packaged foods, an explanation of the suitability of plain tinplate containers is relevant and serves to further explain the corrosion mechanism.

Stage 1: On packing the temperature is high and oxygen in the pack is rapidly reduced and rapid detinning also occurs.

$$Sn \longrightarrow Sn^{2+} + 2e\text{-}, \quad \text{anode}$$
$$O_2 + 2H_2 + 4e^- \longrightarrow 4OH^-, \quad \text{cathode}$$
$$H^+ + e^- \longrightarrow [H], \quad \text{hydrogen absorption by the steel}$$

Stage 2: After a few weeks much of the oxygen will have been consumed and less 'vigorous' cathodic reactions occur, e.g. reduction of natural or synthetic dye-stuffs in the product or traces of nitrate. This results in lower cathodic 'driving current' and hence more controlled detinning.

Stage 3: Progressively further reduction reactions will begin to cease leaving only the small cathodic reaction concerning the generation of atomic hydrogen. The latter will occur slowly and further detinning will be limited by the rate of hydrogen absorption by the steel.

It is this sequence of events which renders the plain tinplate container suitable for many acid products. If, however, the tincoating weight is too low to afford adequate protection for the underlying steel and/or there is an excess of oxygen, then corrosive failure can occur even after Stages 1 and 2 have been reached. These failures are the result of iron dissolution and hydrogen evolution, the latter giving rise to hydrogen swells.

$$Fe \longrightarrow Fe^{2+} + 2e^- \quad \text{at the anode}$$
$$2H^+ + 2e^- \longrightarrow H_2 \quad \text{at the cathode}$$

Anodic dissolution ultimately may give rise to perforation and the ingress of oxygen which greatly affects the rate of corrosion. Leaking product will then attack other containers externally until catastrophic/cascade corrosion of stored packs occurs.

Corrosion limiting packs As a general rule, packages presenting large areas of anode and small areas of cathode do not rapidly corrode because there is little cathodic reaction/current to drive the corrosion reactions. In some case the anodic reaction is limited by the nature of the corrosion product. A typical example of the latter is the formation of tin sulphides in meat and fish packs where the insoluble deposit of tin sulphide limits further attack at the corrosion site. Similar situations occur with iron sulphide.

Tin-free steel Tinplate cans and ends are generally regarded as 'safe-systems' as far as most food products are concerned. The replacement of the tinplate end by a lacquered tin-free steel end generally produces no problems provided the lacquer system has adequate integrity. Tin-free steel (TFS) cannot be used in an unlacquered condition because of its lack of resistance to corrosion and because its hard surface produces unacceptably high tool wear.

Aluminium In some countries, foods such as fish in oil and paté, are packed in all aluminium containers with extremely satisfactory results. By and large, products containing brine are to be avoided as such products have produced rapid and dramatic corrosion of containers. This corrosion has exhibited itself in the form of container and/or end perforation in many cases within 24 h.

Corrosion has also led to the bleaching of red coloured fruits, e.g. fruit salad and fruit cocktail packs, as a result of traces of aluminium, exposed via imperfections of the lacquer coating, dissolving in the product.

Testing for corrosion performance Clearly when we come to pack the product in a particular container specification we expect a long shelf-life. A thorough understanding of the mechanisms by which corrosion occurs is valuable in producing suitable container specifications and removes the risk of unexpected failure with new products.

Container and lacquer specification are tested exhaustively. Most researchers would agree that there is no substitute for packing actual products for realistic periods of time but storage at elevated temperatures is a reliable guide and provides a good indication of the likelihood of success or failure. Chemical reactions (ionic) occur more rapidly at elevated temperatures but specifically cathodic reactions involving diffusion occur significantly more rapidly.

Constantly the researcher seeks improved and instrumented means of predicting pack performance and of understanding the complex reaction which can occur between container and contents. Methods for studying container corrosion include:

- Optical and electron microscopy supplemented by X-ray analysis
- Atomic absorption spectroscopy for (trace) dissolved metal analysis
- X-ray diffraction for plate and corrosion product identification
- Auger electron, X-ray photo-electron and infrared spectroscopy can be

particularly helpful in the study of pitting corrosion and general surface analysis
- Various electrochemical techniques such as ac impedance for predicting the protective qualities of lacquer systems in various substrates

5.10 Recycling

Technical, economic and political factors influence the extent to which materials can be or are recycled. The political factors relating to materials and energy conservation, whilst important, are considered outside the scope of this text.

5.10.1 *Technical factors*

In the case of aluminium and steel (in its various forms) as used in the can-making industry, there is little doubt that both metals can be recycled from a technical standpoint. Each can be collected and remelted to provide a re-usable product.

The sources of scrap are essentially two-fold:

- In-plant can and end making scrap in the form of clean or at worse lacquered skeletal waste resulting from stamping of discs (DWI or DRD bodies and from end manufacture)
- Recovery from the used container stream (domestic waste and collection schemes)

The situation for steel is different from that for aluminium; the latter can be simply remelted and, with minor addition, re-used for can-making. In the case of steel, the presence of tin and the aluminium, for example from easy open ends associated with steel beverage cans, necessitate certain modifications to the recycling process. Tin per se is not a problem in that it can be removed, normally by solution in caustic soda, or diluted with other sources of tin-free scrap. Aluminium is certainly a problem if tin recovery is contemplated in that it reacts violently with caustic solutions. Steel- and aluminium-based scrap can be readily separated magnetically and the normal arrangement is to pulverise the scrap prior to separation.

5.10.2 *Economics*

A significant factor in the economics of recycling is the market value of the metal since the price per ton provides a motive for recovery. On a tonnage basis, aluminium is more valuable whilst the higher density of steel presents advantages in terms of the volume of material to be transported. Further advantage to steel can recycling occurs when the market price for tin is high, unfortunately this market price fluctuates.

Pressures for the conservation of materials and energy sources and for the reduction of land-fill sites, will provide a further driving force for recycling. Recycling schemes for both steel and aluminium are well established in various parts of the world. In 1988, 30% of all tinplate sold in Europe was recovered and recycled; this equates to some 10 billion cans. In the United States and Australia around 50–60% of all aluminium beverage cans are recycled.

References

1. Bodsworth, C. and Bell, H.B., *Physical Chemistry of Iron and Steel Manufacture*, Longmans, London, 1972.
2. McGannon, H.E., ed., *Making, Shaping and Treating of Steel*, United States Steel Corporation.
3. Hoare, W.E., Hedges, E.S. and Barry, B.T., *The Technology of Tinplate*, Edward Arnold, London, 1965.
4. Proceedings of the 1st International Tinplate Conference, London.
5. US Patent No. 155272, US Steel Corporation and European Patent No. EP 0-043-182.
6. Nippon Steel D.A.P. Blackplate.
7. *Printing Ink Manual*, W. Heffer, Cambridge.
8. Pilley, K.P., *Lacquers, Varnishes and Coatings for Food Cans and for the Metal Decorating Industry*, Arthur Holden Inks Ltd. (now ICI Coatings), 1977.
9. *Metal Decorating from Start to Finishes*, N.M.D.A., U.S.A.
10. Food Additives and Contaminants Committee Report on the Review of Metals in Canned Foods (FAC/REP/38), HMSO, London, 1983.
11. Jowitt, F.W., S.M.E. Conference, Chicago, IL, 1986.

6 Packaging of heat preserved foods in glass containers

J. BETTISON

6.1 History

The term 'glass' is used to denote a range of substances made from different materials. Some, like obsidian, are natural glasses produced by volcanic action, whereas others are man-made. The common feature of all these substances is that they are supercooled liquids, i.e. liquids which have solidified without undergoing any significant structural change.

Present evidence suggests that the first man-made glass was produced around 5000 years ago, probably in Egypt. Glass vessels, capable of holding liquids, and which are more than 3500 years old, have certainly been found there. However, the technique of glass-blowing, the logical method for the production of containers suitable for the storage of foods and beverages, appears to have been introduced during the 1st century BC, probably in Syria, where the 'glass' produced at that time was particularly suited for the purpose.

Blowing the glass into moulds in order to achieve consistency of shape and capacity followed soon after, and in those parts of the world under either Syrian or Roman influence, it is not unreasonable to believe that during the 1st and 2nd centuries AD, narrow-necked blow-moulded glass bottles were in common circulation. In fact, the manufacture of bottles and jars continued in the time-honoured fashion of mouth-blowing into water-cooled moulds from the 1st century BC until late into the 19th century.

Glass containers have been used commercially for the preservation of food products by heat for almost two centuries. Indeed, the first processed food containers which were used by the Frenchman, Nicholas Appert, late in the 18th century, were made of glass. Although Appert developed his technique of heat preservation in 1791, he kept it secret until 1809, when it was offered to the French Government in response to an appeal for a method of food preservation capable of providing packaged food for use by Napoleon's forces in Eastern Europe. Appert was paid 12,000 Francs (about £500 in the money of the day) in return for supplying, at his own expense, 200 printed copies of his 'method of Appertisation'.

Since those early days, when natural corks were used to close the glass jars, industry has introduced a variety of closures, with the most significant developments occurring during the past 25 years.

6.2 Properties of glass

(1) As a container material for the whole range of heat preserved foods, glass is perhaps the perfect choice; it is inert, impervious to gases, odours and flavours, and resistant to chemical attack, certainly by chemicals likely to be found in foods.

(2) Glass containers are normally transparent, and this allows the consumer the opportunity to examine the product before purchase (although this does offer a potential disadvantage with certain light-sensitive food products).

(3) Although glass is regarded as a fragile material, a newly formed glass fibre will support a weight twice as much as an equivalent steel fibre. Unfortunately, glass in the form in which it is normally found in real life, is about one thousandth of its theoretical strength; because glass has the structure and properties of a liquid, any defect within it will spread unhindered throughout its mass and thereby reduce its strength. Fragility must therefore be considered as a disadvantage possessed by the conventional glass containers used for the packaging of food products.

(4) Glass is a poor conductor of heat, and sudden changes of temperature, normally regarded as meaning changes in excess of 60–65 °C, in container glass may cause potentially dangerous stresses due to thermal shock; however, sudden heating, which puts the glass surface under compression, is less dangerous than sudden cooling, which puts the surface in tension. It is also important to note, with the increasing trend towards convenience meals and less home food preparation, that the glass container is suitable for use in microwave ovens, once any metal closure has been removed.

This factor, regarding possible glass breakage due to thermal shock, does demand that more care be taken during the heat processing of glass food containers when compared to, say, metal cans, although not necessarily so compared to the newer plastic-based processable containers.

(5) One final disadvantage for glass containers is their weight compared to alternative packaging materials, but developments are under way to reduce this; a recently introduced 'lightweight' glass jar was some 30% lighter than the existing jar, yet gave similar performance with regard to abuse resistance. Computer-aided design (CAD) is able to contribute to the efficient design of glass containers and to the most economic distribution of the glass itself within the container.

6.3 Glass nomenclature

The glass jar is made up of a number of features, the most important of which are shown in Figure 6.1. The hermetic seal, essential for containers to be used for food products, is created between the closure and the sealing surface of the jar itself. This sealing surface has no mould lines across it, being moulded from a one-piece top ring; mould lines are permitted across the threads

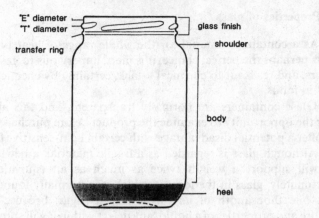

Figure 6.1   Glass food container.

themselves. Dimensionally, the 'E' and 'T' diameters are important, since they control the efficiency with which the caps are picked up by the jars in the capping machine and the degree to which the cap is securely held on its lugs.

6.4   Glass container manufacture

Glass containers for use with heat processed foods are normally manufactured from clear, or flint, glass. For some particular branded products, where a traditional 'stoneware' image is being maintained, jars may be opal (by the addition of traces of fluorine) or otherwise opaque. If required, green or amber glass may be produced by the addition of trace amounts of chromium and iron (green), or iron, sulphur and carbon (amber).

Typically, glass containers for food are manufactured from the following raw materials:

Silica sand	45%
Soda ash	13%
Limestone	10%
Minor ingredients	2%
Cullet	30%

The 'minor ingredients' component may be made up of feldspar, calumite, selenium, etc. which act principally as decolourisers, neutralising impurities such as iron. 'Cullet' is sorted and cleaned glass, made up of containers rejected on site or recycled glass collected from bottle banks. These components are mixed together to form the 'batch', which is then fed into one end of a large furnace, where it is heated to 1500 °C. Molten glass is drawn out at the other end and fed to the container-forming machines (see Figure 6.2).

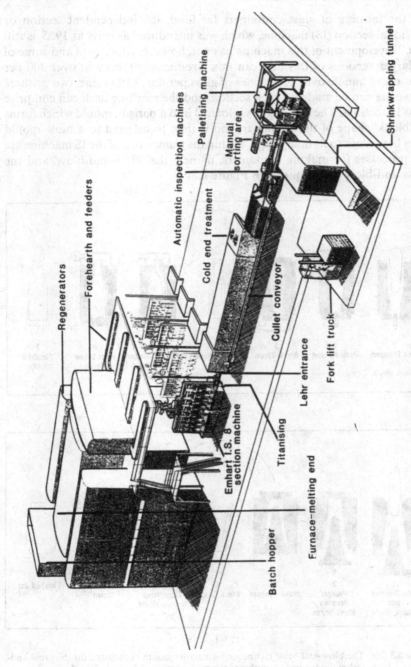

Figure 6.2 A typical furnace (courtesy of United Glass Ltd.).

Regenerators

Forehearth and feeders

Automatic inspection machines

Palletising machine

Manual sorting area

Shrinkwrapping tunnel

Cold end treatment

Cullet conveyor

Fork lift truck

Lehr entrance

Titanising

Emhart I.S. 8 section machine

Furnace—melting end

Batch hopper

6.4.1 *Container-forming machines*

For the forming of glass containers for food, the independent section or individual section (IS) machine, which was introduced as early as 1925, is still used. Development of this machine has not, however, stood still and some of the larger versions of this unit can now produce containers at over 300 per minute and handle over 100 tonnes of glass per day. Either one, two or three containers can be made on each section, and the machine itself can comprise up to 10 sections. The gob of glass is loaded into a parison mould which forms the 'blank' shape of the container, and is then transferred to a blow mould being blown out to its final shape. Within the framework of the IS machine are two processes for making containers, namely the 'blow-and-blow' and the 'press-and-blow' techniques (see Figure 6.3).

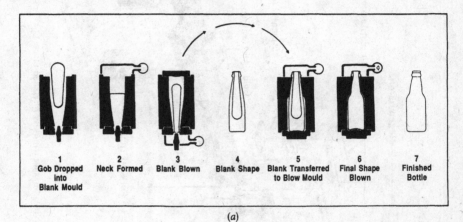

(a)

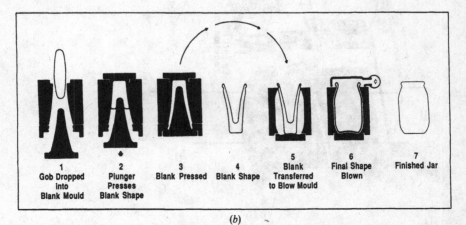

(b)

Figure 6.3 (a) The blow-and-blow technique for narrow mouth containers; (b) the press-and-blow technique for wide mouth containers (courtesy of United Glass Ltd).

With the 'blow-and-blow' technique, which is used for narrow-necked bottles, the gob of glass is dropped into the parison mould and air pressure on the base of the gob enables the neck to be completely formed at the other end of the mould. Air is then blown through this neck to form the embryo body of the container. In the second stage, this partially formed bottle is transferred by its neck to the bottle mould, where the final shape is obtained by air pressure applied again through the neck of the bottle.

With the 'press-and-blow' technique, the gob of glass again drops into a parison mould, but a plunger is then driven into the glass, thereby forming the body cavity and pressing the glass into the required neck finish at one and the same time. This 'press' stage is followed by a final 'blow' stage, as described above. The 'press-and-blow' technique produces containers with a higher degree of dimensional control than that associated with the 'blow-and-blow' technique and, for this reason, increasing use is being made of the 'press-and-blow' technique, even for narrow-necked containers, when high dimensional accuracy is demanded.

Trends in glass-forming are aiming towards increasing output and improving the consistency of quality of the containers produced. In this respect, more and more containers are now being manufactured by 'double gobbing', in which two gobs are formed in the feeder and droped into double cavity moulds; 'treble gobbing' is also being employed on a few units producing small containers. Electronic control of IS machines in place of mechanical control devices permits more rapid change-over from one style of container to another. Needless to say, mould-making is an art, and moulds are extremely expensive. It used to be common practice to coat the moulds with oil every 20 min, leading to loss of production and the possibility of manufacturing a number of rejects, but now the moulds are being plated following manufacture in order to eliminate this need.

6.4.2 *Annealing lehr*

As the containers leave the forming machine, they are still at a temperature of approx. 650 °C, and if allowed to cool too rapidly, they would develop internal stresses and strains and be more susceptible to impact damage. For this reason, they are rapidly conveyed to a temperature-controlled tunnel or 'lehr', through which the containers slowly pass and where they are slowly reheated and then cooled at a predetermined rate.

6.4.3 *Surface treatment*

The trend towards the lightweighting of glass containers has necessitated the increasing use of glass surface treatments in order to maintain the overall 'strength' of the containers, by permitting their smooth flow through packing lines and improving their abrasion resistance. Almost all glass containers now receive some form of external surface treatment during their passage from the bottle-forming machines to the packing station.

The first stage, or 'hot-end' treatment, is applied on the conveyor carrying the containers from the forming machine to the annealing lehr; normally it consists of spraying the hot bottles or jars with a vapour of organic titanium or inorganic tin compounds, thereby producing a very thin layer of the metal within the surface of the glass. Such a treatment is considered to double the strength of the container.

The second stage, or 'cold-end' treatment, is applied to the cooled, annealed containers at the exit of the lehr, and normally consists of an organic compound such as oleic acid, which increases the lubricity of the containers and enables them to move efficiently through high speed filling lines.

Application of surface treatment materials must, however, be carefully controlled, since excessive levels of certain materials, if allowed to come into contact with the neck ring finish on the containers, do have an adverse effect on capping performance, removal torques and the tendency towards rusting of the lugs of the caps. It is common practice to apply cold-end treatments as an under belt spray in order to avoid such contact. It is important for the glass-maker to be in close contact both with the food manufacturer and the cap supplier, so that such potential problem areas are discussed and resolved at an early stage.

6.4.4   Quality control and inspection

After emerging from the annealing lehr, each container passes through a series of automatic inspection devices which are designed to reject weak containers and those which do not meet certain dimensional and quality criteria such as internal neck diameter, sealing surface continuity and the presence of cracks or crizzles.

Quality control personnel are also involved in the monitoring of raw materials, and full dimensional and visual checks on the finished containers prior to despatch. Generally, a quality specification is agreed between the glassmaker and the user, so that routine incoming quality inspection by the food manufacturer is reduced to the minimum. Glass container manufacturers have set themselves up to achieve BS5750 accreditation for their operations.

6.4.5   Design of glass containers

Sophisticated container design facilities incorporating CAD techniques are now available, but there is one general comment which is of relevance whatever the type of closure or container shape decided upon. All packaging material suppliers are continually looking at ways of safely reducing the specification of their materials in order to reduce costs. Significant advances have been made in lightweighting glass containers, and closure manufacturers are being encouraged to follow a similar path of development.

It is therefore necessary to ensure that the design of the glass container itself offers the greatest protection from impact damage, both between the containers and between their closures. The designer should take this factor into account and he should not create a design on which, for example, the shoulders of the jar are so narrow that cap-to-cap contact will occur, since such contact between the skirts of the closure on a high speed packing line could lead to disturbance of the cap, with the consequent reduction in closure integrity and vacuum retention.

6.4.6   *Glass neck ring finishes*

For every type of closure used on glass containers for food, there is an appropriate neck ring finish (Figure 6.4). However, the range of finishes is not as wide as the range of closures available. The glassmakers and closure manufacturers work very closely together in this area and, having decided upon the glass container required (i.e. capacity, shape) and the type of closure to use with it (i.e. diameter, appearance), the appropriate neck ring finish will be selected. Further consideration is given to this feature in Section 6.6.

6.5   Glass recycling: the bottle bank scheme

As many people and many sectors of society became increasingly aware of the apparent waste of used glass containers, and with the increasing awareness of the importance of recycling and energy conservation activities within the United Kingdom in particular and the EEC in general, the Glass Manufacturers Federation (GMF) launched a bottle recycling scheme in 1977 called the 'bottle bank'.

The bottle bank is concerned with the recovery and recycling of glass containers of all types except those for which an economic delivery/collection service already exists or could be created, for example the home milk delivery scheme which is still operated in the United Kingdom and where the average milk bottle is re-used 24 times.

The response of the public in the United Kingdom has been positive, and this initiative is seen as a further major step forward in the creation of industry-wide appreciation of the environmental forces that now encompass the packaging worlds.

When the bottle banks are full, the cullet is normally transferred temporarily to a local storage area until a sufficiently large load (20 tonnes) has accumulated for shipment to one of the glass recycling centres at the major glassworks around the United Kingdom (Figure 6.5).

There is a 'Quality Specification for Waste Glass' which sets out guide-lines

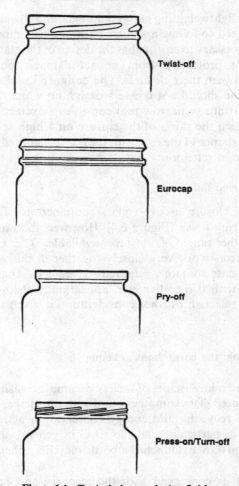

Figure 6.4   Typical glass neck ring finishes.

for the contractors. On arrival at such a centre, the cullet is weighed, inspected and loaded into the reception hopper of the treatment unit itself. Overhead magnets remove all iron-containing items, whilst china, bricks, stones and wrong coloured bottles are removed by hand.

The glass is then broken down by an impact crusher in order to release metal and plastic bands and closures; it passes over a series of vibratory screens which allow the glass particles to drop through, whereas paper or light plastic/metal pieces (especially aluminium) and corks etc. are sucked up by giant vacuum nozzles.

The cleaned glass particles, which are less than 25 mm in size, pass beneath a metal detector which sweeps metal particles (and a small amount of

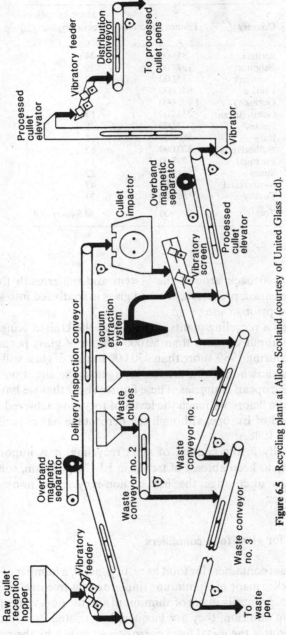

Figure 6.5 Recycling plant at Alloa, Scotland (courtesy of United Glass Ltd).

Table 6.1   Glass recycling in Europe in 1986 (courtesy of British Glass).

Country	Tonnes collected	Share of national consumption (%)
Austria	85 000	44
Belgium	127 000	39
Denmark	35 000	32
France	646 000	26
Germany	1 102 000	37
Great Britain	233 000	13
Ireland	7 000	8
Italy	580 000	38
Netherlands	320 000	62
Portugal	29 000	14
Spain	261 000	22
Switzerland	140 000	47
Turkey	34 000	27
Total	3 599 000	30.5 (average)

surrounding glass) back around the system and underneath the overhead magnets. The cleaned, processed, crushed glass is finally fed into storage silos where is it held prior to use.

Individual glass recycling plants now exist in the United Kingdom with a potential throughput of more than 60 000 tonnes of glass per annum. It is estimated that during 1989, more than 250 000 tonnes of glass will be recycled, but the figures shown in Table 6.1 for 1986 show the situation throughout certain other European countries. These figures show that we have still much progress to make before we reach the levels of recycling achieved in countries on the mainland of Europe, although the percentage has risen from 13% in 1986 to 16% in 1988.

Not specifically in the context of glass recycling, it is important for all packaging users to keep abreast of trends in EEC legislation, some of which are designed to discourage the use of non-recyclable, non-environment friendly packages.

6.6   Closures for glass food containers

Closures for glass containers for food have, in general, a number of similarities. Normally, they are made of aluminium, tinplate or tin-free steel (TFS), with the venting closures primarily made of aluminium and the non-venting closures of tinplate/TFS; in addition, they are lacquered or coated both internally and externally to protect the metal from corrosion or attack by the product itself, and to enhance the sales appeal of the package. Finally, they are usually

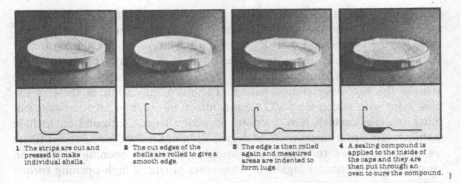

1 The strips are cut and pressed to make individual shells.

2 The cut edges of the shells are rolled to give a smooth edge.

3 The edge is then rolled again and measured areas are indented to form lugs.

4 A sealing compound is applied to the inside of the caps and they are then put through an oven to cure the compound.

Figure 6.6 Method of manufacture of metal caps (courtesy of CMB Packaging).

provided internally with a ring of lining compound which in some closures also extends down the skirt of the closure to form both a top seal and a side seal.

6.6.1 *Closure manufacture*

The manufacture of such metal closures basically consists of the operations shown in Figure 6.6. The coils of metal (tinplate, TFS or aluminium) are cut into sheets of convenient size for handling through the printing and lacquering machines. These sheets are then passed a number of times through printing presses and coating units which apply printing inks and varnishes to the outside of the sheet and product-resistant lacquers to its inner surface. The choice of lacquer(s) used depends on the type of food product to be packed. Between each of these operations, the sheets are passed through a heated oven, which is able to cure the coated materials and bake them on to the surface of the plate, often at temperatures in excess of 215 °C. All the above operations are performed on the flat sheets of metal prior to any forming or shaping of the closure itself.

The printed and lacquered sheets are then slit into strips ready for feeding into stamping presses which cut them into disks of appropriate size to form the caps. These disks are pressed and curled into shape and the liquid lining compound, which creates the sealing gasket around the periphery of the cap, is then injected into the shell of the cap whilst rapidly rotating. This rotary action serves to flow the lining compound into the channel of the cap, although with some types of closure it is also necessary to press or mould the compound to encourage it to arrive at the appropriate position. A final heat treatment is given in an oven in order to cure and set the compound, often at a temperature in excess of 210 °C.

Perhaps one of the most critical factors in ensuring optimum performance of closures on glass jars is the formulation of the lining compound itself. At the present time, the vast majority of lining compounds used in closures for food are based on thermo-plastic PVC resins. A number of different grades of PVC resin are used, depending on the type of application, e.g. pasteurisation or sterilisation, for which the closure is designed. However, apart from the PVC resin, a number of other ingredients have to be incorporated in order to achieve the necessary characteristics. The most important of these is the plasticiser with which the PVC resin is mixed in order to form a paste, but other materials which may need to be added to the formulation include stabilisers (in order to prevent the degradation of the PVC on heating), fillers (in order to soften the lining compound and reduce opening torques on closures) and lubricants (again to counteract potential high opening torque situations).

Quality control   Throughout all the various manufacturing processes, regular quality control checks are carried out on the raw materials and the closures themselves, in order to ensure that the finished closures conform to the critical standards demanded for high speed operation in packing plants. Such checks cover the basic properties of the tinplate, TFS or aluminium as received from the suppliers; examination of other raw materials such as printing inks, lacquers and lining compounds; measurement of the applied film weights and curing conditions (times and temperatures) of the lacquers and coatings used; the performance of the applied lining compound, and dimensional tolerances of the closures themselves. Finally an application trial and a processing treatment, which involves the use of an autoclave, may be carried out.

6.6.2   Closure classification

Closures for glass containers for food may be divided into two broad categories, namely the venting closures and the non-venting closures. In the simplest of terms, a venting closure may be defined as a closure which is simply crimped on to the jar and which then permits, by controlled venting, the escape of air entrapped in the jar during subsequent heat treatment. With the onset of the cooling phase, a vacuum is created in the jar, this vacuum pulling the closure down and thereby resealing it on the neck of the jar.

By contrast, a non-venting closure is applied to the jar under vacuum, thereby giving an immediate hermetic seal, and the closure is maintained in this fully sealed condition throughout subsequent heat treatments by the application of a controlled overpressure or counterpressure to more than balance the internal pressure developed by the product within the jar itself.

Venting closures (see Figure 6.7)   The Omina cap and the Pano range of closures are typical venting closures. These closures, which are normally made of aluminium, are not hermetically sealed on the jars when applied, the application of the cap being achieved by clinching the rim and/or skirt of the

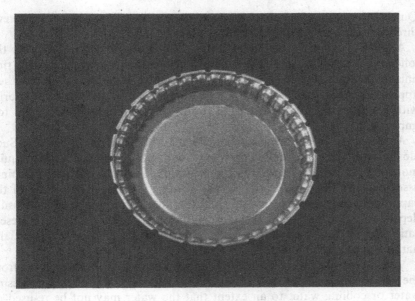

Figure 6.7   Typical venting closure (53 mm Eurocap).

cap around the neck of the jar. Under normal closing machine setting conditions, these closures will start to vent when the internal pressure in the jar reaches $0.2\,kg/cm^2$ (3 psi). Critical machine setting is essential in order to give acceptable cap venting conditions.

During pasteurisation, when the product is normally heated up to 85–90 °C, the air in the headspace, and sometimes also some of the product or the liquid governing the product, escapes from the jar. During cooling, a vacuum begins to be created in the jar and as soon as the internal pressure falls below the external pressure surrounding the jars, the closure is hermetically sealed.

During sterilisation, where temperatures up to 121°C are normally used, the internal pressure built up within the jar could be high and it is therefore necessary to apply a counterpressure or overpressure within the retort in order to control the rate and degree of venting. If the counterpressure used is too low, there will be excessive product losses from the jar and possible deformation of the aluminium closure itself, which could prevent it from resealing correctly during cooling. If the counterpressure used is too high, the rate of venting may be too slow to give an acceptable final vacuum level and again possible deformation of the closure itself may occur.

There are basically three types of Pano closure available, namely the Pano N (normal), the Pano Quick-Lip and the Pano T. All three are made of aluminium and all are suitable for both pasteurisation and sterilisation heat treatments. They differ essentially in the means of removing them from the jars; the Pano N requires some artificial assistance of lever to prise it off, the Pano

Quick-Lip has a lip for pulling it off and the Pano T closure is used on screw-threaded jars and is removed by unscrewing.

A general summary of the advantages of venting closures indicates the following. Venting closures generally require less sophisticated glass neck ring finishes and therefore, in general, they are easier to manufacture. Due to their application and venting characteristics, they may also operate satisfactorily with overfilled jars. Thus, accurate control of filling conditions is not too important.

Among their potential disadvantages may be listed the following. Such closures are generally applied on rotary multi-head machines which require more regular maintenance than the straight-through capping machines associated with the most popular non-venting closures. The setting of the capping machines is very critical since the final performance of the closure depends on the ability of the closure to vent during heating, yet reseal satisfactorily during cooling without permitting the ingress of potentially harmful spoilage organisms.

Venting of overfilled jars can lead to the expulsion of product or liquid from the jars themselves and this could lead to contamination of the processing water or cooling water to an extent that the water may not be re-used or recirculated. With certain products such as pickled vegetables packed in vinegar or acetic acid, the acidity of the water could be increased to such an extent as to lead to accelerated corrosion of equipment and the water circulation system.

Such closures may require external assistance or leverage for removal and, in general, they cannot be reclosed.

Non-venting closures (Figure 6.8)   Typical of this group may be mentioned the Whitecap style of closure. These are normally made of tinplate or TFS and are hermetically sealed on to the jars. The application of the cap is normally achieved under vacuum by the injection of steam into the headspace of the jars and surrounding the closures with steam during application. The vacuum created in the jar headspace is sufficient to prevent the venting of the closure during normal pasteurising conditions at less than 100 °C, although a calculated overpressure or counterpressure is necessary during sterilisation of such jars at temperatures in excess of 100 °C. The appropriate counterpressure is easily calculated using a pressure diagram, from which the maximum internal pressure generated in the jars is established by knowledge of the product filling temperature, the level of vacuum created during closing, the headspace in the jars and the sterilisation temperature being used.

The Whitecap Vapour-Vacuum capping system has now been in operation for more than 50 years and a high degree of sophistication has been established. Within the capping system, provision is made for the steam treatment of the closures themselves just prior to application, which serves

Figure 6.8 Typical non-venting closures (51PT and 63RTO).

both to pre-sterilise the inner surface of the closure and to soften the lining compound in order to facilitate the creation of the hermetic seal.

Such closures are normally applied to multi-start threaded glass finishes permitting easy removal and reclosure of the jars. The majority of these closures are fitted with lugs which are locked beneath the threads of the jar neck by twisting the closures. Such closures are known as Whitecap RTO (regular twist off), MTO (medium twist off), DTO (deep twist off) closures, and they may be applied to jars at speeds in excess of 400 jars per minute on relatively simple straight-line Whitecap capping machines. They are commonly available in diameters from 30 to 110 mm. However, for even higher speed application (in excess of 1200 jars per minute is already being achieved in Europe), the Whitecap PT (press on–twist off) is available in a range of diameters from 27 to 70 mm. This closure differs from the twist off type in that it is applied by simply pressing it down on to the jar without any rotary action, since it is not fitted with lugs, because of the placement of the lining compound down the skirt of the closure itself. This lining compound takes up the shape of the threads on the glass neck ring during pasteurisation or sterilisation and it is possible to screw the closure off.

Whitecap style closures, when used for sterilised products, are not normally subjected to temperatures above 125 °C; maximum total retort pressure of 2.45 kg/cm^2 should not be exceeded.

The advantages of non-venting closures are as follows. Non-venting closures ensure that the product is maintained in optimum condition throughout its life, free from oxidation and free from the possible ingress of harmful spoilage organisms. The fact that the jar is sealed under vacuum permits the detection and removal of improperly sealed or damaged jars prior to processing by the use of dud detection equipment in the packing line immediately after the capping machine. An easy open and reclosure facility is provided.

The potential disadvantages of this system are the need for accurate filling and sealing conditions in order to ensure consistent vacuum in the jar, and control of overpressure throughout the sterilising operations in order to prevent displacement of the caps or excessive disturbance of the lining compound/glass seal created during application. Consistency in opening torque levels also depends on the above factors, as well as requiring attention by the glass supplier to surface treatments used.

The ratio of venting closures to non-venting closures used for pasteurised and processed food products in Europe is probably around 1:10. These figures do not take into account the use of the traditional Phoenix-type closures which are still being used, though to a lesser and lesser extent, in Eastern Europe or the still widespread use of non-venting crowns on bottled products such as sterilised milk.

6.6.3 Recent developments in closures

More recently, flexible aluminium foil laminated lids have been heat-sealed on to specially treated glass finishes as a closure system for certain food products such as hot-filled jams and preserves. Because the heat-sealing operation normally requires a dwell time in excess of 1 s, multi-lane sealing machines are usually used in order to achieve even relatively low outputs. Such jars are also normally provided with a plastic overcap which serves as a reclosure after the primary seal has been removed, adding a further capping operation.

Plastic caps have been used for bottled liquid products such as sauces and vinegars, and some pickled products in wide-mouth jars, without serious difficulties. Basically, there are two types of plastics closures, both of which are manufactured from modified polypropylene resins. The simpler type of cap does not contain a lining compound but relies on features such as flexible tabs or annular rings moulded into the cap during manufacture to create a seal on the glass finish. The second generation plastic caps are produced with PVC-based lining compounds which give improved closure performance on the glass jars. Of course, the main difficulty in the manufacture of such caps is in curing the PVC compound without melting or otherwise distorting the plastic cap itself by the application of heat. A number of techniques are being investigated in order to produce an undistorted cap with a satisfactorily cured lining compound, including the use of a combination of hot air and microwave energy, or even redesigning the closure itself to allow it to distort to the correct shape during heating; other more sophisticated techniques are under development. Such developments are still in an early stage and at present plastic closures are confined to hot-filled or pasteurised products only.

The application of such caps under conventional steam closing conditions also has to be controlled carefully in order to avoid heat distortion of caps. However, judging by the pace of development, it is likely that we will see plastic closures capable of withstanding sterilisation treatments in closed retorts within the next decade, probably also incorporating a tamper-evident feature.

There are also a number of metal/plastic combination closures which offer the following potential advantages: By using a metal disk, identical to that on a conventional all-metal cap, across the neck of the jar, both the seal achieved on the glass sealing surface and the performance achieved in contact with the product itself, should be identical to that given by the conventional metal closure. By using a plastic skirt down the thread-finish of the jar, one should be able to achieve convenience of opening, elimination of external rusting or corrosion and if required incorporate a tamper-evident feature.

At present, the range of such closures is limited by, amongst other things, the heat resistance of the plastic materials being used for the skirt, but it is almost certain that future developments will extend the sophistication and range of application for such closures.

Reference should also be made to the developments being undertaken by the 'traditional' metal cap manufacturers. They are continually searching for ways of reducing the cost of their products, or at least for ways of minimising the impact of the apparently ever-increasing costs of raw materials such as tinplate, lacquering and coating materials, energy, etc. To this end, the manufacturers of tinplate caps have further developed their manufacturing techniques (e.g. introduction of energy saving ultraviolet curing of printing inks, use of water-based lacquers, etc.) and are now well advanced in introducing closures made from tin-free steel (TFS), where a very thin coating of chromium is used to replace the heavier coating of tin with a slight reduction in overall costs.

The appearance of TFS is not as bright as that of tinplate, but very little resistance to such a change on aesthetic grounds has been encountered. The extension of TFS, manufactured by a double reduction technique, is also at an advanced stage of development; this harder plate permits a tamper-evident button feature.

Tamper-proofing and tamper-evidence   The excellent reclosure characteristics of twist off and similar closures, whilst offering convenience to the end-user, also create a potential problem area in that this reclosure facility could be abused, e.g. by children in supermarkets. The tamper-evident button feature, already well established, e.g. on the Whitecap PT cap, is now being extended to other closures. When the vacuum is released from such caps, the button at the centre of the cap pops up, and food manufacturers using such closures normally incorporate into their design some instruction to reject containers if found in that condition.

Other tamper-proofing devices are available including the application of a paper band across the skirt of the cap and the neck of the jar, the incorporation of a plastic collar or clip across the shoulder of the jar, or the use of a PVC heat-shrinkable seal across the cap and the jar shoulder. The ultimate technique is to enclose the entire jar and closure in a transparent heat-shrinkable film.

Reference has been made above to the relative ease of incorporation of a tamper-evident band feature on the skirt of plastic or metal/plastic composite closures.

6.7 Handling of glass food containers

Having selected the appropriate glass container and the appropriate closure system, there are five areas in the packing plant and distribution system which are most important in ensuring that the jars reach the consumer in perfect condition:

- Filling conditions
- Capping conditions
- General jar handling on the line itself
- Processing conditions and equipment used
- Subsequent warehouse storage of the filled stock

Because of the wide predominance of the vacuum sealed non-venting style closure in the glass food container market, these are assumed to be used throughout this section; where deviations are required to accommodate the venting-style closure, they are mentioned.

6.7.1 Filling conditions

When filling glass containers with food products, it is important to ensure the following.

(1) Clean filling conditions must be maintained in order to prevent contamination of the sealing surface of the jar. Such contamination could give problems with certain food products; for example contamination with fish or meat products which contain fibrous matter could prevent the creation of a perfect seal between glass and closure, and contamination with products containing starches or milk solids have been shown to increase the opening torque required to remove the caps.

(2) The product filling temperature must be maintained within a consistent range. The product temperature used has a significant influence on the build-up of pressure within the jar during high temperature processing. It is recommended that a minimum filling temperature of 65 °C is used for processed foods, although higher temperatures are to be preferred and lower

temperatures can be accepted with appropriate modification of the over-pressure used during processing.

(3) The filling level must be consistent, and the headspace volume within the filled jars must be adequate to prevent excessive pressure build-up during processing. It is recommended that the nominal headspace within the brimful jars should be not less than 6% of the brimful capacity of the container. Lower headspaces could present difficulties of cap displacement or the use of excessively high over-pressures during processing.

(4) A flat product surface, without projections above the jar rim and with the minimum of occluded air, must be achieved. With certain products, e.g. fruit and vegetables, entrapped air may be kept to a minimum by blanching them before filling, whereas for other products, such as solid meats, a vacuum mixing/filling technique may be required.

With venting-style closures, items (1) and (2) are still important, but the achievement of a headspace is less important, and a flat product surface is not important at all.

6.7.2 *Capping conditions* (Figure 6.9)

Capping machines for the closing of glass containers with non-venting closures are available in a range of operating speeds from 10 to 1500 jars per minute. Apart from the slowest laboratory style units, all are essentially similar in operation.

It is most important that the setting-up and operating procedures of the

Figure 6.9 CMB 2000 capping machine.

equipment are adhered to when closing glass containers filled with food products.

Because of the in-line operating principle of these units, changeover from one size or type of closure to another, or one type of container to another, is achieved quickly. The caps are orientated in the correct condition to seat above the jar neck in a rotating hopper equipped with pick-up pins or magnets; the caps are then fed through a steam-heated chute to the point in the capper where they are picked up by the jar, steam is injected into the jar headspace and the cap is finally pressed or turned onto the jar neck ring finish by overhead sealing belts.

It is necessary to control the performance of the capper, and the most important factors in assessing this performance are as follows.

(1) A cold water vacuum check, which entails filling jars with cold water to a nominal 10–12 mm headspace, closing the jars and then measuring the internal vacuum, should achieve a minimum of 50 cmHg.

(2) A vacuum check on product filled jars should normally be within the range of 20–40 cm Hg, depending on product temperature and headspace being achieved, and consistency in such vacuum levels is an important parameter.

(3) A visual check should be made on the closed jars to establish that the closure is sealing correctly on the glass finish, and that a satisfactory impression is achieved in the lining compound.

(4) A security check should be made by measuring the tightness of application used with lug-type closures.

The pressure and flow of steam to the capping machine, which both evacuates the air from the headspace of the jars and softens the lining compound in the caps just prior to application, is important. However, when handling certain food products which may not be able to accept occasional droplets of water in the headspace from the condensed steam, either superheated steam at $> 140\,°C$ (which is dry) or a cold closing technique using a mechanical vacuum capper may be utilised.

Venting style closures are normally applied on rotary machines. Since such closures are not applied under vacuum and normally do not incorporate a reclosure feature, checks (1) (2) and (4) do not apply.

6.7.3   General jar handling

As with all containers for food, filled and closed jars should be handled carefully throughout all stages from the moment the jars are closed to the moment they are finally sold in the retail store.

Many of the potential handling problems which can occur in the packing plant could be overcome by attention to line layout at an early stage in setting up the factory. Potential damage points should be examined in the existing line and modifications made where necessary, e.g. to the side guide-rails along conveyors and to the equipment being used to load retort crates or baskets, in order to avoid container breakage or damage to the caps.

It is particularly important to avoid cap-to-cap contact, which could disturb the hermetic seal between the glass sealing surface and the lining compound within the closure. Conveyor speeds, dead plates etc. should be set up in such a way that the containers do not make heavy contact with each other.

For the vacuum closed non-venting closures, the use of in-line dud detectors will immediately eliminate incorrectly sealed or damaged packs at the exit of the capping machine. It is also possible to carry out a second such check on non-venting closures after processing/cooling and at that stage a similar check may also be carried out on venting closures.

6.7.4 *Processing conditions*

Apart from the vital part played by the processing conditions used to ensure that adequate sterility is achieved in the jars to give a safe product with an acceptable shelf-life, care has also to be taken to ensure that no unacceptable stresses are placed on the closure during the processing operation, i.e. stresses which could reduce the security of the seal created during the capping operation. Different conditions exist and different procedures are required depending upon the type of heat treatment being given to the non-venting style closures, as follows.

Hot filling   The hot filling technique is used for a wide range of acid products such as fruit juices, preserves (conserves) and condiments (sauces, ketchups and relishes), and the only special consideration to be taken into account when using glass containers for such products is the necessity to avoid thermal shock failure, i.e. glass fracture which could arise when an instantaneous temperature rise in excess of 55–60 °C occurs.

Avoidance of such failure may be achieved by preheating the glass container prior to filling, either by subjecting it to a temperature of 50–60 °C, or by passing the containers beneath infrared heaters. The latter technique is required when filling certain products which rely upon the concentration of some ingredient, e.g. sugar or vinegar, to guarantee continued preservation after heat treatment, since the presence of droplets of water remaining in the container after rinsing could locally dilute this component and result in microbial growth and spoilage.

As the product is not heated to a temperature higher than its filling temperature, there is no increase in the internal pressure built up within the container, and no precautions such as the application of overpressure on the closures are required.

Pasteurisation   The two commonest types of pasteurisation treatment given are to:

- Cold-filled products such as pickled vegetables (onions, gherkins, etc.) which may be heated from their 20–25 °C filling temperature to their 65–75 °C pasteurisation temperature.
- Hot-filled products, such as fruit juices, where a product filled at, say, 80 °C is given a 10 min pasteurisation treatment at 85–90 °C.

With the cold-filled products, it is important to achieve as high an internal vacuum level at the time of cap application as possible, so that any internal pressure built up with the increase in product temperature resulting from the pasteurisation treatment is minimal; most of the non-venting closures will accept some internal pressure before they leak.

With the hot-filled products, the increase in temperature, and thus the increase in internal pressure, is so low that the achievement of vacuum levels as low as 20 cmHg will be sufficient to maintain a vacuum within the containers at all times. Again, thermal shock breakage of the glass containers must be avoided by preheating them prior to filling.

For the venting closures, the cold-filling technique is most widely practised, since this facilitates the production of a good internal vacuum level on cooling.

Sterilisation   During heat processing at temperatures in excess of 100 °C (normally in the range 115–121 °C), the actual internal pressure built up in a glass container closed with a non-venting closure is governed by a combination of filling and capping conditions and the sterilisation temperature being used. The term 'overpressure' is used to signify the total pressure in the sterilising chamber, and is the sum of the steam pressure which is required to achieve the processing temperature and the pressure of the superimposed compressed air. Glass jars closed with non-venting closures are normally sterilised under steam-heated water, although it is also possible to sterilise in steam/air in purpose-built retorts (e.g. Lagarde), in water recirculating overpressure retorts (e.g. Steriflow) and certain hydrostatic sterilisers (e.g. Hunister and overpressure Stork).

When using conventional static vertical retorts for under water processing, it is important to ensure that all the jars in the retort are covered by at least 15 cm of water, and that there is a headspace of at least 20 cm between the top of the water level and the top of the retort. Some retorts are capable of rotating the crates of glass containers and, under controlled conditions, this will present no problems, other than the selection of closures capable of performing satisfactorily under such stringent conditions.

There are limitations to the overpressure which some caps will withstand, and it is not therefore a simple matter of using a high overpressure for all situations. By reference to a pressure calculator chart (Figure 6.10), it is possible to calculate the maximum internal pressure which will be developed in the containers during sterilisation, knowing the following:

- The internal vacuum in product filled containers

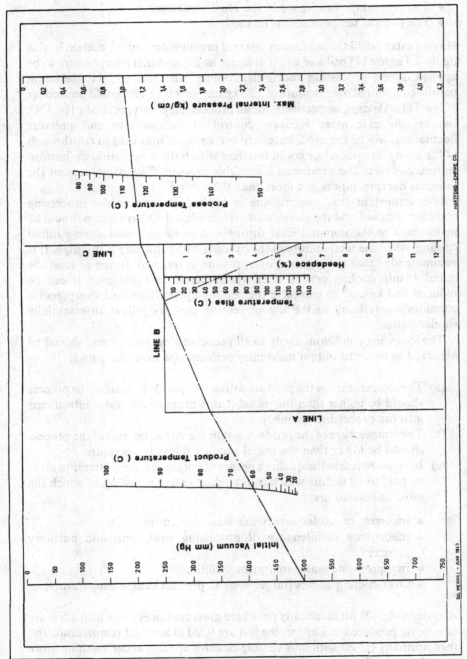

Figure 6.10 Nomogram for pressure developed in a sealed glass container (courtesy of Holmegaards Glasvaerk).

- The product filling temperature
- The percentage headspace in the filled containers
- The process temperature to be used

Having established the maximum internal pressure developed, add on to this figure $0.3\,kg/cm^2$ (5 psi) and use that figure as the nominal overpressure to be applied during processing and initial cooling. Alternatively, the maximum permitted nominal overpressure of $2.25\,kg/cm^2$ or $1.9\,kg/cm^2$ could be used for PT or TO style caps, respectively, for all process temperatures above $104.4\,°C$, but in this case more accurate control of overpressure and pressure fluctuations will be required. Excessive overpressure may lead to cut-through of the lining compound or could interfere with button operation on button-feature closures. The maximum acceptable pressure fluctuation around the nominal overpressure is not more than $0.2\,kg/cm^2$.

It is important that overpressure is applied as soon as the processing chamber is closed and the steam/water is introduced. Overpressure should be maintained at the nominal level throughout processing and during initial cooling, when the temperature of the product within the jars is still high. It is recommended that the nominal overpressure is retained during at least the initial 15 min cooling period, after which, based on experience, it can be reduced to $1\,kg/cm^2$ to complete the cooling cycle. The retort overpressure conditions pertaining to the use of venting closures follow an essentially similar pattern.

The following conditions apply to all processing operations and should be observed in order to obtain maximum performance from the packs:

(a) The temperature of the product within the jars at the start of the process should be higher than the initial temperature of the water introduced into the processing chamber.

(b) The temperature of the product within the jars at the end of the process should be lower than the initial product filling temperature.

(c) In order to retain the quality appearance of the closures, attention must be paid to all factors which could induce external rusting, of which the most important are:

- Incorrect or inadequate boiler water treatment
- Inadequate maintenance of processing equipment and ancillary services
- Inappropriate water treatment facilities
- Over-cooling of jars (target $= 40\,°C$ product centre temperature)

Aseptic filling With aseptically packaged glass containers, of which a few are now being produced in Europe, the jars are filled at ambient temperature and then normally closed with non-venting closures applied under vacuum. Since no subsequent heat treatment is given to the container after capping, the vacuum level achieved on the capping machine will be maintained throughout the shelf-life of the container.

6.8 Quality control, warehousing and storage

Routine checks should be carried out on processed jars in order to confirm satisfactory production performance, and the following tests are common:

- Lining compound impression must be consistent across the jar sealing surface and show no signs of thinning or cut-through
- Vacuum checked after cooling will normally be in the range 30–45 cmHg for pasteurised products and over 45 cmHg for sterilised products
- Opening torques normally checked 24 h after processing/cooling, and a further check may be performed 2–3 weeks later
- Dud detector rejects should be checked to establish the reason for failure, and necessary corrective action taken if under plant control

In order to minimise the disturbance of the caps and to allow the lining compound to settle down, excessive top loading must be avoided. This can be achieved by paying attention to the palletisation and stacking of the finished pallets, and the following recommendations are made for container handling throughout the period they remain in the warehouse prior to final distribution:

- Where jars are packed into shrinkwraps for distribution, it is recommended that microflute board is used for the shrinkwrapped trays, since this has been found to give the highest degree of resistance to cap abuse caused by excessive top loading.
- The pallets used must be in good condition and palletised loads must be lowered gently in position to minimise uneven distribution on the pallets beneath. It is also recommended that divider pads be placed between each stacked pallet to reduce still further the uneven distribution caused by single-faced pallets; such divider pads should be made either from 12 mm solid wood or 16 mm hardboard, chipboard or Masonite board.
- The stacking height of the pallets is significant with regard to possible top loading abuse of the caps. As long as the above recommendations are followed, there would appear to be no significant increase in top loading abuse on pallets stacked three high in warehouses for long term storage, whereas four high stacking of pallets must be carefully controlled. Stacking of palletised jars more than four pallets high is not recommended.

Acknowledgements

The author wishes to thank Robin Moorby of United Glass Ltd, for his valued assistance and also colleagues at CMB Packaging Technology PLC for their help and for permission to prepare this text.

7 Packaging of heat preserved foods in plastic containers

P.J.G. PROFFIT

7.1 Introduction

The last twenty years have seen a pronounced change in lifestyles. Factors such as more working mothers and wives, greater numbers of elderly persons and single persons living on their own have brought about the need for more convenient meal preparations. Factors such as the increase in extracurricular activities for children and adults have meant a need for staggered meal times. The increase in dual income families in itself has meant changes in lifestyle patterns.

Microwave cooking fits ideally into this modern and developing framework of society. In fact in 1988 sales of microwave ovens had penetrated 90% of households in the United States and 45% in the United Kingdom. These changes have all happened during the 1980s and similar patterns are happening in other countries such as Japan, Australia and mainland Europe. It is expected that these demographic trends will continue well into the next century, thereby creating a high demand for convenience foods from what is its current state of infancy.

The microwave oven has brought with it an opportunity to develop a unique form of food packaging. This new container concept not only provides the primary packaging functions but also a receptacle in which the food product can be first microwaved and then placed directly onto the table. Plastic materials are ideal for these new retortable packages because not only do they provide the properties required for retort processing but they also facilitate the transmission of microwave energy. Other consumer market advantages and perceptions which contribute to the suitability of retortable plastics packaging are:

- Unbreakability
- Hygienic image
- Portability

Figure 7.1 Multi-layer barrier plastics containers from Japan.

7.2 Market trends and growth

The market position in the United States for retortable plastic packaging leads Europe by 3–5 years. The United Kingdom and Germany lead in Europe, probably by a similar margin over other European countries. Japan also has a mature retortable plastic packaging market although many of the Japanese applications tend to be medium-acid and therefore do not always use full retort sterilisation temperatures and conditions.

The convenience food market is still in its infancy. We have seen interest growing from the early 1980s and will see long steady growth well into the next century as demographic trends continue in their current direction.

Retortable plastic packaging shares the convenience food market with other prepared foods, namely chilled and frozen foods. For ready meals eaten at home, shelf stability, combined with easy microwave reheating provides retortable plastics packaging with some convenience benefits over chilled or frozen preparations, but in areas where food is eaten away from home, such as in the office or in vending outlets, retortable plastic packaging offers unique benefits over current alternative forms of packaging.

These new packaging concepts usually cost more than their traditional counterparts such as metal or glass containers used in retort applications. It must be understood that in these consumer driven markets high added value

Figure 7.2   Heat processed shelf stable barrier plastics containers aimed directly at microwave reheating in the home.

processed foods, such as ready meals, must justify these higher costs. Indeed, in the case of chilled foods these costs have to be balanced against the expense of operating chilled storage and distribution.

7.3   The properties of plastic packaging materials

The physical properties of plastic materials (polymers) show some major differences when compared to traditional materials such as metal or glass. Polymers are usually softer and more flexible. In technical terms their modulus of rigidity reduces with an increase in temperature and this varies for different polymers. Polymers also display 'creep' or movement under stress particularly at elevated temperatures. This phenomenon does, for instance, mean that conditions used in closing containers by double seaming have to be optimised for plastics in retort processing because of the forces exerted and the elevated temperatures of processing.

The process used for forming polymers will also affect strength and creep properties. Polyethylene terephthalate (PET) is a classic example of this. In its amorphous phase, PET is malleable. It has very poor creep properties and is quite soft once its temperature increases above 65 °C. If PET is oriented it becomes very strong and has excellent creep properties. In this condition it may be used in high tension applications such as rope manufacture. If a further heat setting process is applied similar to that employed by oriented PET film manufacture, it not only remains strong in tension but also withstands high temperatures up to 140 °C. If amorphous PET is crystallised in its unoriented form it has even higher thermal stability (up to 230 °C). This technology, known as CPET, is exploited in the manufacture of dual ovenable trays for frozen foods.

Polymers do not provide complete barriers to gases. This means that oxygen will permeate through the wall of a container into the product. The rate of permeation will depend upon the polymer or polymers employed and is known as the polymer's 'barrier properties'. Polymers also allow moisture to permeate and again this varies with different polymers.

7.3.1 *Requirements for polymer structures*

The two key requirements for retortable plastics containers are:

- Strength to withstand retort conditions at temperatures up to 135 °C
- The appropriate barrier requirement to oxygen for the type of food to be packed and its required shelf-life

Further to these requirements comes the strength to meet abuse resistance, stacking, transport and storage requirements. The container has also to have the ability to accept the closure system which for retort will probably be applied by double-seaming or heat-sealing techniques. Food being packaged must not pick up taint and odour from the polymer and of course the polymer must meet appropriate food packaging regulations for the relevant retort regime employed during processing.

Having achieved all the physical requirements, the price of the polymer must make it possible for containers to be produced economically and competitively.

7.3.2 *Combining polymers for optimum performance*

To enable all requirements for retortable containers to be met, it is common to combine various polymers together to provide optimum cost/barrier/thermal performance. These combinations will usually comprise of a main polymer or structural polymer that has good thermal and strength properties to meet retort requirements with a barrier polymer that provides the structure or multi-layer system sufficient resistance to the permeation of oxygen for the type of food to be packed and its required shelf-life. These two polymers will be combined in a multi-layer structure using various techniques of processing.

Depending on whether a natural bond exists between the structural and barrier polymers, a third polymer may be introduced in the form of a 'tie' layer which glues the barrier polymer to the structural polymer. It is normal in such multi-layer structures to bury the barrier polymer in the structural polymer to form the three or five layer systems seen in Figure 7.3.

In processing techniques such as extrusion blow moulding or thermo-forming, the production line produces an inherent amount of trim waste

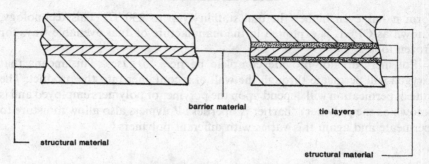

Figure 7.3 Typical three and five layer barrier structures used in food packaging.

which, for reasons of economy, is reprocessed back into an additional reclaim layer within the multi-layer structure. If this is the case, the five layer system becomes six layers usually taking the form shown in Figure 7.4. This asymmetric structure may be symmetrical when two reclaim layers are present.

7.3.3 Structural polymers

Criteria for structural polymers are that they should be economically priced, as they generally constitute 80–90% of the total raw material content and that they provide a good modulus of rigidity at the required retort temperature. These polymers must, as mentioned above, provide suitable properties to meet other mechanical requirements of the pack. The closure system used usually engages with the structural polymer and so this polymer has to be capable of being double-seamed or heat-sealed, depending on the type of closure used.

Polypropylene   The most popular polymer used to day for retortable plastics containers is polypropylene. This polymer is capable of withstanding retort conditions as well as providing good mechanical strength at room tempera-

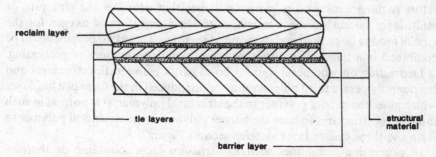

Figure 7.4 Typical six layer structure incorporating a reclaim layer.

tures. Its one drawback, however, is that its low temperature impact strength is poor. It is also not an easy material to thermo-form, although the injection moulding process presents no problems. Polypropylene is produced as a homo-polymer, or in a co-polymer form with polyethylene. The latter variation, although adding about 10% to the price, does however produce improved low temperature impact performance. Both varieties are used for manufacture of retortable plastic packaging.

Polypropylene is historically priced close to polyethylene which places it at around 70% of the price of ordinary PET and close to half the price of CPET.

The oxygen barrier of polypropylene is very low and for food packaging applications, this polymer is nearly always combined with a high barrier material in a multi-layer structure. Polypropylene does, however, have a good barrier to moisture at ambient temperatures. Moisture barrier properties, however, do reduce considerably at typical retort temperatures.

Certain nucleated grades of polypropylene may be used to produce containers showing good 'contact clarity'. This property enables the product to be visible when the inner surface is wetted. Improved clarity is achieved if this polymer is used in an injection moulding process where improved surface finish of containers is possible.

There are, unfortunately, no commonly used barrier polymers that adhere to polypropylene without the use of tie layers; such adhesive polymers providing sufficient strength to withstand the vigour of the retort process are readily available from several suppliers.

Polyethylene terephthalate (PET) PET is currently the second most important polymer used in the manufacture of retortable plastic packaging. PET, as mentioned earlier, has poor strength at elevated temperatures when in its amorphous phase. Processing PET for high thermal stability is carried out by compounding a nucleation material into the raw material; when formed in a hot mould at temperatures of 170 °C, the PET structure goes crystalline which gives the final container a high degree of strength at temperatures up to 230 °C. These thermal properties are superior to those of polypropylene as can be seen from the curves showing the modulus of rigidity in Figure 7.5.

This form of PET is commonly known as CPET and the raw material is usually sold as a compound ready for the sheet extrusion process. Although all PET will crystallise at a peak crystallisation rate at around 170 °C, the added nucleant does speed up this process to produce more economical mould cooling times.

CPET displays moderate barrier properties to oxygen without the addition of a barrier layer. This barrier level makes it a suitable polymer for some of the less oxygen sensitive foods. At this point in time, there are no known commercial high barrier CPET containers on the market.

CPET raw materials are approximately 30% more expensive than equivalent PP barrier structures, but the less sophisticated mono-layer production

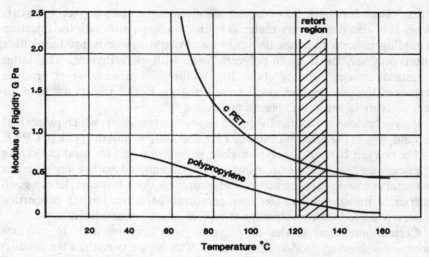

Figure 7.5 The reduction of strength of polypropylene and CPET with temperature.

process does offset this. Indeed the higher strength of PET at retort temperatures should enable weight reduction of up to 10% to be achieved for the equivalent duty container.

Polycarbonate   Polycarbonate is a polymer that provides excellent mechanical strength, both at room temperature and retort temperatures. It is also a relatively easy polymer to process but its price as a raw material is likely to be around twice that of CPET, making it expensive for container production. In addition to this, it possesses extremely poor barrier properties and would thus have to be used in a multi-layer structure to produce a practical container. It is not currently used in any popular retortable plastic packaging applications.

7.3.4   Barrier polymers

There are three important high oxygen barrier polymers that are used for food packaging. These polymers are generally only suitable for use within a multi-layer structure as their mechanical properties are usually not very good. The cost of barrier properties dictates their economical use and it is important that the manufacturing system employed produces good control of the barrier layer with an even thickness as this will have a high impact on both container performance and cost of final product.

Ethylvinylalcohol (EVOH)   EVOH is important as a polymer that provides a high resistance to the permeation of oxygen when in its dry state. Because of its ease of processing it is the most popular of barrier polymers. The drawback of

EVOH is that its barrier properties reduce as its moisture content increases. This does present a reduction of barrier after retorting due to moisture pick-up.

If a PP/EVOH structure is retorted, moisture passes through the polypropylene (which at high temperatures has a low barrier to moisture) and into the EVOH thus significantly reducing its oxygen barrier. After cooling the container, the moisture in the EVOH now becomes trapped by the restored high moisture barrier of polypropylene. It then takes two or three weeks for the EVOH to dry out and restore its high barrier properties. It is a requirement therefore, for higher levels of EVOH to be used in retortable packs to achieve the equivalent barrier levels of non-retorted containers. Retort time and grade of EVOH will also vary this effect, as will its position in the structure relative to the outer layer of the container.

The ethylene content of the EVOH is varied to tailor its processing characteristics. The lower the ethylene content, the better the barrier; but as higher draw and lower temperatures may be required during forming of sheets into containers then the ethylene content is increased to enable the barrier layer to be stretched without breaking. In processes where material is formed at melt conditions, the temperatures employed during forming are higher; it is therefore possible to use the higher barrier grades of EVOH. This is a drawback for reheat solid phase forming processes where lower barrier grades are employed for the equivalent draw ratios.

Extrusion characteristics of EVOH make it a relatively easy resin to process provided the correct extrusion screw is employed and the extrusion die system allows no hold-up points for material during its passage through to the die lips. It is usual practice to dry EVOH prior to extrusion as, although it usually arrives dry and packed in sealed bags, it will pick up moisture from the atmosphere if it is left for a period of time. Consequently, if moist, bubbling in the barrier layer will occur as this moisture boils off during processing.

Polyvinylidene chloride (PVDC)   PVDC is also an important barrier polymer which has been used in film applications for food packaging for many years. Although PVDC has slightly lower barrier properties than EVOH in its dry state it is not affected by moisture in the same way as EVOH and therefore gives overall an improved barrier performance when used in retort applications.

PVDC is, however, a very difficult polymer to process efficiently, requiring specially constructed extrusion tooling made from duro-nickel alloy. This highly degradable polymer requires a carefully controlled manufacturing process otherwise scrap rates can offset its advantages. Reprocessing of multi-layer scrap is also difficult, requiring special extrusion systems designed together with reclaim additives to minimise degradation.

Polyamide (PA)   Polyamide (PA) and its derivatives are used to provide an

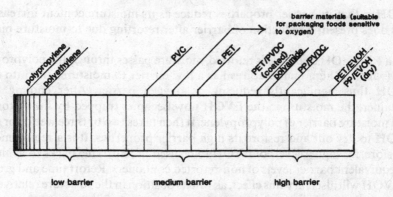

Figure 7.6   The ranking of oxygen barrier properties of various polymers.

oxygen barrier within multi-layer structures. Although lower barrier properties compared to EVOH and PVDC are attainable, structures with PA have been popular in Japan for many years. High barrier amorphous polyamide is popular for use as a barrier material with PET, its main advantage being that melt temperatures are compatible and some natural bonding exists between these two polymers. PA/PET is highly suitable for co-injection moulding processes, the two material structure being much simpler to manufacture than if a tie layer were employed.

Polyamides' barrier to oxygen is not, however, affected by moisture and as such it has potential for retortable plastics.

Figures 7.6 and 7.7 illustrate the oxygen barrier and thermal stabilities of polymers.

7.3.5   Tie layers

These specialist resins are made from various polymers or polymer compounds to produce a highly polar material providing the necessary bonding properties. The essential requirement is that they bond to both the structural and barrier resins maintaining the strength of bond at retort temperatures. They also need to be easily processable in the virgin state as well as within a mixture of the multi-layer components. Typically these polymers are polyethylenes maleic anhydride graft modified or acrylic acid or ester graft modified.

7.3.6   Reclaim considerations

Where the plastics process employed involves the trimming of components producing an inherent trim waste fraction, for reasons of economy, this is normally reprocessed back into the structure as an additional reclaim layer. The primary polymers used to make a multi-layer structure have to be capable of being combined in a reclaim extruder to produce a homogenous melt with

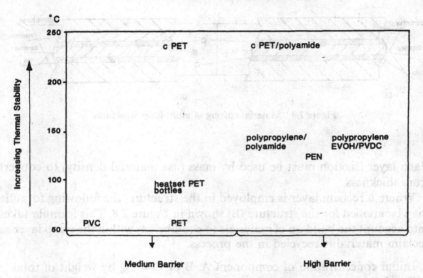

Figure 7.7   The positioning of various polymer barriers and thermal stabilities.

no degradation. This can sometimes cause difficulties, for instance, with combinations such as polypropylene and PA where the temperature required to melt the polyamide would cause the polypropylene to degrade, resulting in potential taint and odour problems.

A similar difficulty occurs with PET/EVOH where the EVOH will degrade at PET melt temperatures. Polypropylene mixes well with EVOH provided steps are taken by the raw material producer to minimise the catalyst residues in the polypropylene. It is, therefore, essential that compatibility of melt temperatures and mix of materials is taken into account if reclaim is involved in manufacture.

If such incompatibility exists, material additives called 'compatibilisers' may be employed to allow reprocessing of the scrap. This relatively new technology is becoming increasingly important where the plastics industry is increasingly continuing to produce highly optimised multi-material products.

7.4   Material costing of multi-layer structures

Material costing of simple structures where two or three raw materials are employed is relatively straightforward. Each component is taken as a percentage by mass of the total and this, together with the raw material cost, is calculated as shown below for the structure (A) shown in Figure 7.8.

$$
\text{Average multi-layer price per tonne} = \begin{array}{l} (\text{Fraction A} \times \text{price/tonne}) \\ + (\text{Fraction B} \times \text{price/tonne}) \\ + (\text{Fraction C} \times \text{price/tonne}) \end{array} \quad (7.1)
$$

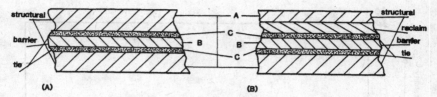

Figure 7.8　Material costing of multi-layer structures.

Each layer fraction must be used by mass (use material density) to convert from thickness.

Where a reclaim layer is employed in the structure, the following formula may be applied for the structure (B) shown in Figure 7.8. This formula takes into account the build up of materials to equilibrium within the scrap layer as reclaim material is recycled in the process.

Initial concentration of component A, B or C = $x\%$ by weight of total
Regrind component　　　　　　　　　　　　= $y\%$ by weight of total

Total percentage ingredient in the whole structure $= \dfrac{x}{1 - (y/100)}$

This calculation is made first and then used to calculate cost as before (Eq. 7.1). Figure 7.9 shows some typical barrier resins or resin combinations giving their approximate price per tonne together with their barrier ranking.

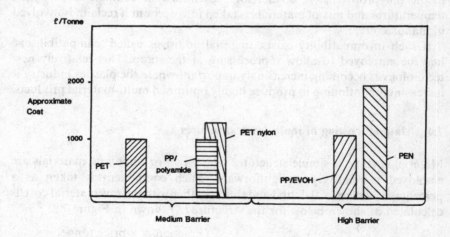

Figure 7.9　Barrier materials and combinations used, ranking their approximate price and barrier properties.

7.5 Container manufacturing technology

Potentially there are a number of processes that may be used for the manufacture of retortable plastic packaging for food applications. These processes may be adapted or designed to combine various polymers to form multi-layer containers. Such processes include:

- Injection or co-injection moulding
- Injection or co-injection blow moulding
- Sheet extrusion and co-extrusion combined with thermo-forming
- Extrusion or co-extrusion blow moulding
- Orientation blow moulding for mono- and multi-layer PET or poly-propylene bottles and jars

The four techniques used to combine polymers to make multi-layer structures are:

- Co-extrusion for sheet film or tube
- Co-injection moulding
- Lamination
- Coating

The first three of these combination techniques are used in various forms today for the manufacture of retortable plastic packaging.

7.5.1 Co-injection moulding

This technology is an extension of conventional injection moulding using two or three injection units working with one mould to produce either a part component or a preform which is later blow-formed using compressed air inside a mould to make a bottle or jar.

The co-injection moulding process (Figure 7.10) uses the injection units to meter materials into cavities in such an order that the barrier material flows through the main structural material to create a multi-layer structure. This process is commercial in the United States using three materials for production of polypropylene-based retortable containers.

The key to success in this process is in the accurate control of material flows, particularly where multi-cavity moulds are employed. Several operations around the world are commercial using PET and polyamide to make bottles or jars but only one company has so far been successful in commercialising a three material multi-cavity system (structural/tie/barrier). Such five layer containers may be seen in supermarkets in the United States.

As mentioned earlier, the injection moulding process may complete the container manufacture or a special version of the machine is employed, designed to convey the injected preform still hot and held on its mould core from the injection station into a blow station where compressed air is used to

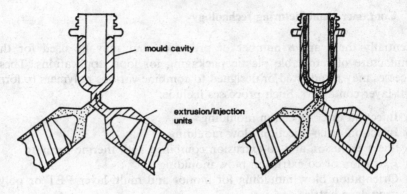

mould cavity

extrusion/injection
units

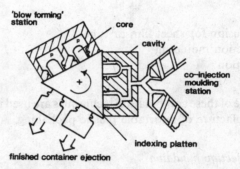

'blow forming'
station

core

cavity

co—injection
moulding
station

indexing platten

finished container ejection

Figure 7.10 The basic principles of co-injection moulding and injection blow moulding. Top left, injection of structural polymer; top right, injection of barrier polymer (p = air pressure).

form the finished container. A variation of this injection-blow process is used to make oriented PET or PP containers.

7.5.2 Co-extrusion

Co-extrusion is the more commonly employed process for combining polymers in the manufacture of retortable plastic packaging. Adaptations of this technique are used to produce tube sections for bottles or film and flat sections again for film or sheet. The tube process may be coupled with a blow moulding stage as shown in Figure 7.11 to manufacture multi-layer bottles widely used in packaging of sauces and relishes. Three or four extruders are used to combine materials into a multi-layer tube. This tube is then transferred in its molten state into a blow moulding machine where compressed air is used to form the final bottle or jar. The component is finally trimmed to remove top and bottom 'flash' which is ground up and fed back into a fourth extruder to form the reclaim layer.

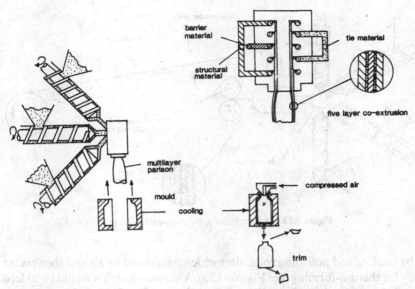

Figure 7.11 The basic principles of co-extrusion blow moulding. Trim waste is usually fed back into the extrusion head to form an additional layer making a six layer structure (p = air pressure).

Sheet co-extrusion uses similar principles to that of tube technology where three of four materials are combined within the extrusion tooling to form a multi-layer sheet. It is usual with sheet processes to form a multi-layer core of extrudate within a 'feed block', prior to spreading this core in a 'coathanger die', to form the final sheet (Figure 7.12). This sheet is then taken away from the

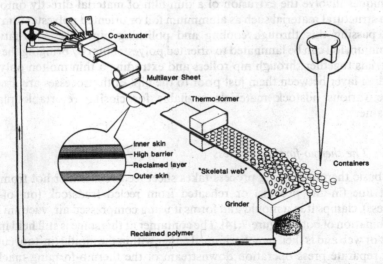

Figure 7.12 Schematic line layout showing sheet co-extrusion and thermo-forming processes.

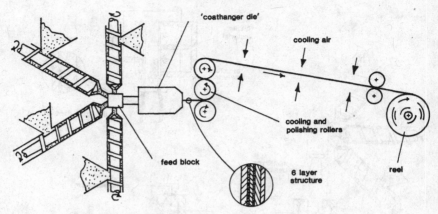

Figure 7.13   Principles of a sheet co-extrusion line.

die by cooling and polishing rolls, after which it is cooled by air and then reeled ready for thermo-forming (see Figure 7.13). A variation of this would be to feed the still hot or molten sheet direct into a thermo-forming machine where it would be formed into containers. This process is known as 'in-line thermo-forming', as distinct from 'out-of-line thermo-forming'.

7.5.3   Lamination technology

Lamination and extrusion coating are important techniques for the manufacture of lidstock materials used in retortable plastic packaging. These techniques involve the extrusion of a thin film of material directly onto the main structural material such as aluminium foil or oriented polyester film and then passing this through cooling and polishing nip rolls. Alternatively aluminium foil can be laminated to oriented polyester film by bringing the two materials together through nip rollers and extruding a thin molten polymer bonding layer between them just prior to the nip. Both processes are used to make various lidstock materials available for closing retortable plastic containers.

7.5.4   The thermo-forming process

The basic thermo-forming process takes sheet feedstock either hot from the sheet line (in-line process) or reheated from reeled feedstock (out-of-line process), clamps it in a mould and forms it using compressed air, vacuum or a combination of both (Figure 7.14). The container at this stage is still held in the sheet or web and is then punch trimmed either within the mould (in-tool cut) or by a separate press operation downstream of the thermo-forming machine (out-of-tool cut). The remaining skeletal waste is usually ground up and fed back into a reclaim layer in the sheet extrusion process.

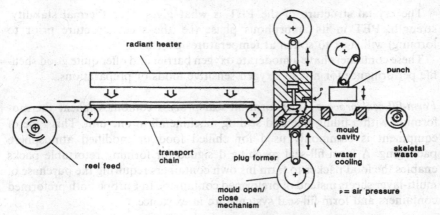

Figure 7.14 The basic reheat thermo-forming process (p = air pressure).

Polypropylene thermo-forming Polypropylene is not an easy material to thermo-form. In its molten state direct from the extruder its low 'melt strength' makes it very difficult to handle into a thermo-forming machine due to sagging problems encountered. Rotary forming processes are employed to overcome these problems but these systems are not readily available from machinery suppliers.

If forming is carried out through a reheat process, two approaches are made. The sheet may be heated to melting point prior to forming, an extremely difficult process to control, again because of melt strength problems. The material may alternatively be formed just below its melting point where handling of hot sheet is simpler. In either case, the key to the process lies in accurate temperature control of the web prior to forming. Polypropylene requires a very narrow temperature tolerance band in the sheet if even wall thickness containers are to result.

Whatever route of manufacture is used, the process has to be designed to minimise internal stresses in the wall of the containers. Any residual stress here will show itself as distortion of container shape after retorting.

Thermo-forming of CPET Thermally stable PET containers are in common use for dual ovenable applications for chilled and frozen foods. Some containers are made using this process for 'retort shelf stable' food applications. Typically these containers known as 'crystallised PET', or CPET are stable at temperatures up to 230 °C.

The CPET process is based on conventional reheat thermo-forming where an extruded PET sheet containing an added nucleant material, is reheated to above its softening temperature and thermo-formed into a hot mould (around 170 °C) where it is held for long enough for a crystalline structure to develop within the PET. The forming is then transferred into a second mould where it is cooled and finally enters an out-of-tool trim press.

The crystal structure of the PET is what gives it its thermal stability/ strength. PET in its amorphous phase (i.e. the sheet structure prior to forming) will start to soften at temperatures over 63 °C.

These containers have a moderate oxygen barrier and offer quite good shelf-life performance for not so oxygen sensitive foods or preparations.

Form-fill-seal process The form-fill-seal process combines reheat thermo-forming with a filling line and sealing machine all in one unit. This type of equipment is commonly used for chilled food or modified atmosphere packaging. A form-fill-seal machine designed for forming retortable packs enables the food packer to form his own containers requiring the purchase of multi-layer sheets instead of preformed containers. In Europe both preformed containers and form-fill-seal systems are in existence.

7.6 Closure systems for retortable plastic containers

7.6.1 *Requirements of closure systems*

A key factor in the marketing requirements for ready meals and other shelf stable preparations is 'true convenience'. This must not be forgotten when it comes to the closure system and here, 'easy openability' is an important feature.

In addition to this requirement, other features necessary for packaging low acid foods are:

- A 'bio tight' seal
- A controllable process that can be operated between known process control limits by the manufacturer
- An economical closing system
- A closure system compatible with the plastics processing tolerances associated with the container manufacturers

For the future of retortable plastic packaging, the closure process must be developed such that all these aspects are taken into account in the final design of equipment employed, be it double-seaming, heat-sealing, ultra-sonic welding etc. It is fair to say that the industry still has a long way to go in understanding and developing the closing process and there exists a great desire for improved process control.

7.6.2 *Headspace oxygen*

The headspace is an important factor to be considered for both pack shelf-life and retort performance. The headspace air entrapped in a closed container using a conventional flat foil sealing system will significantly affect product shelf-life. Headspace air will also effect the retort performance of the container.

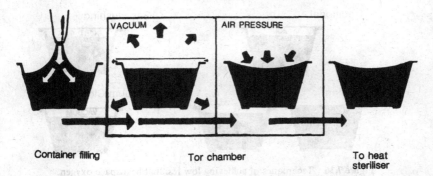

Figure 7.15 CMB TOR process illustrating principles of formation of hydraulically solid pack.

Plastic polymers have lower strength at retort temperatures, considerably lower than traditional glass or metals materials. It is therefore important to protect the containers from excessive stress in the walls brought about by pressure differential between the inside of the pack and the external pressure in the retort during the heating and cooling cycles. It is usual to minimise the effects of container stress by providing an overpressure in the retort vessel (typically 1.75–2.00 bar total at 121°C); but while there remains any gas at all in the headspace of the container, then minimising differential pressure completely will be very difficult.

It is these stresses generated particularly during heating and cooling where temperatures are transient that causes pack distortion. The most straightforward approach to improve this situation is to close the packs in such a way that headspace gas is eliminated or nearly eliminated. By doing this the overpressure can then work against a 'hydraulically solid' pack thus minimising container wall stresses. In this condition, a constant overpressure may be applied from start to finish of the retort cycle. The CMB patented TOR process facilitates the provision of a hydraulically solid pack (Figure 7.15).

Headspace minimisation also provides a pack with minimum oxygen content as well as good contact between wall and contents for heat transfer. A further consideration of using this technique to minimise wall stresses during retort are that lighter weight containers should be achievable over systems where headspaces exist. Common techniques of achieving minimal headspace are (Figure 7.16):

- Use of predished lidding
- Use of steam flush during closing
- Drawing a vacuum on container prior to sealing
- Brimfilling

A predished foil system is designed for the closure to minimise the headspace and perhaps even displace the product to the brim of the container. This technique generally must leave some air in the headspace if the likelihood of

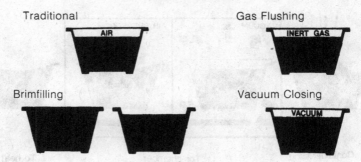

Figure 7.16 Techniques of achieving low residual headspace oxygen.

product overflow and hence flange contamination increasing the potential for seal failure are to be eliminated.

Steam flow or evacuation technique used with flat or dished foils are the most effective systems of minimising oxygen in the headspace, but it must be remembered that the pack design must be such that deflection can occur in the container or the foil to take up the gap between the product and the closure.

Finally brimfilling has similar drawbacks to those using predished closures where accuracy of fill volume becomes critical to avoid flange contamination.

7.6.3 Closure systems

The two closure systems most widely employed with retortable plastic packaging are foil or film laminates and double-seamed ends of the full aperture, easy opening variety. (Figures 7.17, 7.18).

The 'easy open' double-seamed end may be attractive to a food processor because he may already be familiar with double-seaming technology and may be able to use some already existing can closing equipment in his plant. Although double-seaming uses the same basic process as that of metal, some of the techniques, seamer settings and component profiles are quite different to those used for metal to take into account differences in the properties of the plastic flange and container. Foil and film systems have some cost advantages, although some of these are offset by the more expensive sealing process.

In its simplest form heat-sealing consists of clamping a polyethylene container flange together with some polyethylene film using a cold and a heated jaw as shown in Figure 7.19. During the controlled time that the materials are clamped, heat from the jaw penetrates through the film, melting it together with the polymer adjacent to it in the flange. The hot jaw in this process requires a 'non-stick' surface such as PTFE to avoid it bonding to the molten film.

An extension of this principle is used for retortable plastics. The lidding used is usually a laminate whose structural material may include either polyester or

Figure 7.17 UK products showing both heat sealed and double seamed closures.

Figure 7.18 Double seamed containers—USA containers manufactured by American Can
Company.

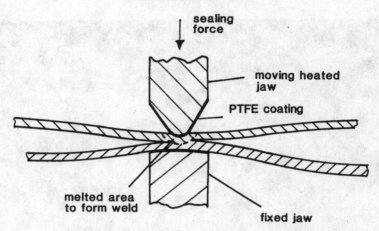

Figure 7.19   The principle of simple heat sealing.

aluminium foil coated with polyester. The heat-sealing material is usually coated to the underside of the lid stock. In this process the heated jaw is set at a temperature which is lower than the melting point of the material it directly contacts (usually polyester or perhaps aluminium), but higher than the melting point of the sealing polymer. The heat then transfers through the substrate and into the interface between sealing layer and container flange thus causing them to melt and bond together (see Figure 7.20). In this type of operation, higher sealing jaw temperatures may be employed to attain shorter sealing cycles.

Foil lidding systems   Foil lidding systems are ideal from the shelf-life viewpoint. The closure in the example of a tray represents a large proportion of the surface area of the whole container and when a polymer lidding system is employed, the additional oxygen transmission through the lid will have a significant effect on shelf-life performance. Foil systems do offer 'total barrier' to oxygen and thus give the best shelf-life performance compared to all-plastic closure systems.

Aluminium is a good material for heat-sealing due to its superior heat conduction properties enabling faster sealing cycles than the equivalent polymer system.

The top surface of the aluminium foil may be laminated with oriented polyester film giving it improved tear strength and puncture resistance as well as covering any pinholes that may exist in the aluminium foil. The underside of the foil is usually coated with a special heat-sealing compound or compatible polymer film which readily seals to the container material but also satisfies the peelable opening characteristics. Much development has occurred in the development of such materials and these lidding systems are today available commercially from several suppliers.

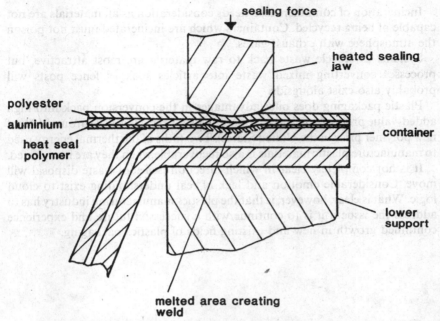

Figure 7.20   The principle of sealing typical lidding for retortable plastic container.

Polymer lidding systems   For various reasons, the all-plastic clear lidding systems have generally more appeal to the consumer enabling the product to be seen through the lid. Barrier properties of the all-plastic lidding are important for good shelf-life performance. Materials such as oriented PET film are used for the substrate to give the necessary strength and heat resistance together with a barrier resin. EVOH, polyamides and PVDC are used to provide this oxygen barrier.

7.7   Environmental issues

In our society today there is a fast growing awareness of the pressures upon all individuals to deal with the problem of waste in an 'environmentally friendly' way, paying attention to emissions, energy preservation as well as conservation of raw material resources. It is short term thinking and irresponsible for the manufacturing industry to enter with new products in new markets without considering all aspects of the product life cycle through to disposal.

Today, the recycle industry is fast growing, developing processes for 'cleaning' used containers back into usable polymer. PET and HDPE systems are examples of this. Responsibility rests with individuals and governments to ensure that raw material is collected and separated in such a way that these industries can be assured of reliable feedstock supplies.

Incineration of containers also needs consideration as all materials are not capable of being recycled. Containers which are incinerated must not poison the atmosphere with exhaust gases.

Systems to recycle waste back to raw material are most attractive, but processes converting mixed waste into articles such as fence posts will probably also exist alongside.

Plastic packaging does offer advantages in the conversion back into high added-value products such as raw material (PET recycle sells at around 40% of new polymer price in the United States). Two fifths of the thermal energy used to manufacture plastic materials is also recoverable when they are incinerated.

It is not completely clear in which direction issues of waste disposal will move. Considerable emotion and lack of real understanding exist to cloud logic. What is clear, however, is that the plastics manufacturing industry has to address the issue if it is to continue with a successful future and experience continued growth in new and existing fields of plastics packaging.

8 Leaker spoilage of foods heat processed in hermetically sealed containers

R.H. THORPE

8.1 Introduction

The microbiological safety and stability of heat processed foods depends on packing appropriately prepared food into containers which can be hermetically sealed, applying an effective heat process to destroy micro-organisms capable of growth in the food and ensuring that the contents are not subsequently re-infected by leakage of micro-organisms into the containers.

This chapter deals with the requirements for conventional canned foods but the reader should appreciate that much of the content applies equally well to other packaging systems and other heat preserved products. The same need to prevent leaker infection into the container exists with aseptically packed products as well as those in-pack heat processed foods packed in glass and plastic containers.

8.2 Leaker infection—the commonest form of spoilage

Gillespy [1], discussing the control of leaker spoilage, suggested that in the early days of the canning industry the expensive hand-made cans seldom leaked and the majority of spoilage was due to inadequate processing. However, Goldblith [2] pointed out that Nicholas Appert, who first published details of his method of preservation in 1810 and is generally acknowledged as the father of canning and heat processing, was aware of the problem of leaking cans some years before publishing the fourth edition of his book in 1831. In the fourth chapter of this edition, Appert gave a guarantee under the heading of 'Conditions on which M. Appert Guarantees His Preserves' in which he pledged to replace blown cans provided defective stocks were returned to him. In the same chapter he provided the first leaker spoilage statistics in that he stated 'it is possible to find one or more out of twenty-five which will need to be thrown into the ocean because of a bad odour coming from a small opening, either in the closure or in the body of the tin. This is a recognised truth by more than fifteen years experience.' He then dismissed this high failure rate, which today would be completely unacceptable, in his next sentence 'This loss is very small compared to those which are encountered with stewing, salting and so many other foods on which the furnisher has no recountability.'

Gillespy [3] stated that infection by leakage in canned vegetables was much too frequent for complacency on the part of the UK industry. He qualified the statement to make it quite clear he was not referring to blown cans, which could be expected to be removed before the point of sale. Thorpe and Everton [4] stated that leaker infection was the commonest form of spoilage encountered in low acid canned foods in the United Kingdom. Data, generally obtained from records of investigations of spoilage outbreaks, have been published by a number of workers.

Subba Rao [5] reported spoilage in Indian canned foods, Kefford and Murrell [6] in Australia, updated by Richardson [7]. Segner [8] provided spoilage data for 1960–1977 in the United States as did Pflug et al. [9] and the National Meat Canners Association [10]. In another survey, Ababouch et al. [11] reported on spoilage in Morocco.

For the United Kingdom, the reader is referred to [1, 3, 12]. No other data giving spoilage figures have been published in the United Kingdom. It should be appreciated that many food processing companies have qualified technical staff who investigate spoilage problems. Therefore, the can manufacturers or independent laboratories may not automatically be consulted, unless the leaker spoilage is believed to be associated with 'out of can manufacturers' specification' of containers and ends.

The spoilage data quoted confirm that, for Australia, India, the United States and the United Kingdom, leaker infection is the commonest form of spoilage encountered by the industry. Although information for other countries is not available, there is no reason to believe that the situation is markedly different. A further 18 countries covering four continents are listed by Stersky et al. [13] in their extensive review of food poisoning incidents associated with leaker spoilage and this supports the belief that leaker infection is widespread and the commonest form of spoilage generally faced by the canning industry.

8.3   The economic cost

The economic cost of leaker spoilage has been reported by Shapton and Hindes [14]. Murrell [15] reported that the financial loss to the Australian industry that year was $A0.38 million.

The 1964 outbreak of typhoid fever in Aberdeen which originated from the consumption of canned corned beef cooled in unchlorinated river water [16, 17] resulted in 515 cases and three deaths [18]. The sales of the particular brand were stopped during the outbreak and the subsequent investigation. More significantly, overall sales of corned beef were depressed in the United Kingdom for a number of years. It has often been stated in conversation that the market took at least 10 years to recover to the pre-1964 sales level. This statement is difficult to substantiate. However, official government statistics show that prior to the outbreak the average household consumption of corned beef for 1959–1963 was 0.7 oz per person per week. This dropped to 0.43 oz in 1964, averaged 0.5 oz over the next 10 years (1965–1974) and subsequently increased to an average of 0.64 oz for 1975–1979. While it is recognised that other factors could have influenced the consumption figures it does appear that the original stagement has some substance.

In a later incident involving corned beef, the trade in 1980 estimated that a Department of Health and Social Security (DHSS) ban of 13 weeks on £3 million worth of 6 oz cans of Argentinian product had cost £160,000 in interest charges alone [19].

The case of botulism in Belgium in 1982 associated with a UK brand of canned salmon packed in Alaska, stopped all sales of $7\frac{3}{4}$ oz cans of salmon of US and later Canadian origin for nearly six months in this country. Trade was also stopped for varying periods in a number of European and Commonwealth countries besides the United States, where nearly 60 million cans from nine factories were recalled [20]. In the United Kingdom over 23 million cans were held in warehouses for the period of investigation. On top of this, further substantial costs were incurred in conducting the special sorting scheme for cans, both in the United Kingdom and North America, required by the DHSS before batches could be released for sale in the United Kingdom. An additional and considerable expenditure was incurred by a national TV advertising campaign aimed at restoring confidence in the product and recovering sales in the United Kingdom. The economic loss resulting from food poisoning incidents, including some involving leaker infection of canned products, was discussed by Todd [21].

8.4 Type of micro-organisms associated with leaker spoilage

When the occurrence of leaker spoilage has to be confirmed, most workers concentrate their investigations on the condition of the container and the source of infection. The typing of organisms involved in the spoilage incident is not always completed except where food poisoning has occurred or is suspected.

Thorpe and Everton [4] stated that cans infected by leakage generally contained non-sporeforming organisms. Bashford and Herbert [22] reported that normally micrococci and the small bacilli gained entry to leaking cans and small gram negative motile organisms were frequently prominent in cultures from spoiled leaking cans. The data published by Kefford and Murrell [6] support the previous statements. However, Pflug et al. [9] considered mixed cultures containing aerobic or anaerobic sporeformers as well as mixed or pure cultures of non-sporing aerobic micro-organisms as typical leaker spoilage organisms.

Thorpe and Everton [4] stated that infection by bacterial spores was very rare. But more recently, the first author has investigated two leaker spoilage cases where it was subsequently shown that spores of mesophilic bacilli were the infecting agents. Bean (1982, private communication) has also investigated several leaker spoilage incidents in which, by inference, the data indicated spores of mesophilic organisms were the cause of infection.

Gillespy [1] stated that infection by yeasts is rare, even in acid products, unless there are gross faults in the can seams. Yeasts were only recovered from one of the seven cases of spoilage in fruit packs reported by Kefford and Murrell [6] and Richardson [7]. Growth of moulds in leaking cans is invariably associated with holed containers. While some holes occur as a result of a critical seaming defect, for example a misassembly, or as a result of pinholes or fractures in the tinplate or transit damage, the vast majority are currently caused by case opening knives used in supermarkets. Stersky et al. [13] listed nine incidents where illness occurred and the cans showed case-cutter damage.

8.5   Food poisoning arising from leaker infection

When related to the annual production figures, the frequency of infection by food poisoning organisms is extremely low. Bashford and Herbert [22] stated that the annual consumption in the United Kingdom exceeded 6 000 000 000 cans of food packed at home and overseas. The Ministry of Health, Public Health Laboratory Service reports for the four years 1961–1964 showed a total of 54 outbreaks of food poisoning orginating from canned foods, less than one outbreak for every 400 000 000 cans consumed. They contended that this indicated a standard of safety unequalled by any comparable foodstuff. Gilbert et al. [23] quoted UK figures for 1978. Only five recorded incidents of food poisoning associated with canned foods had occurred in that year. Against this, the home production and imports amounted to over 2.2 million tonnes net can contents. They stated that even if there was a large element of under-reporting of cases of food poisoning or food-borne illness, canned foods were remarkably safe when the volume of consumption was taken into account.

The same confidence in the safety of canned foods was expressed by Stersky et al. [13] who reviewed 154 incidents of food poisoning associated with post-process leakage in canned foods covering the period 1921–1979. A wide range of low acid products (meats, fish, vegetables, poultry, recipe and milk-based packs) produced in 21 countries were involved. The majority of the food poisoning incidents occurred in the United Kingdom and with few exceptions were associated with imported products. A further 17.2% of the incidents listed were reported as occurring in Canada since 1972.

Leaker infection involving Salmonella typhi caused 6 (3.9%) of the 154 incidents. Other organisms involved were: other Salmonella species, 9 (5.8%); Clostridium botulinum type E, 3 (2.0%); Clostridium perfringens, 3 (2.0%);

miscellaneous and undetermined agents, 33 (21.4%). However, by far the most common organism was *Staphylococcus aureus* accounting for 100 (64.9%) of the incidents. The very high proportion of outbreaks (111 out of 131) caused by staphylococcal poisoning is again shown by Gilbert *et al.* [23], who tabulated the numbers of general and family outbreaks of food poisoning associated with freshly opened canned foods for the period 1929–1980 by infecting micro-organisms. Cockburn [24], discussing food poisoning in the United Kingdom for the period 1949–1958, pointed out the much higher rate of staphylococcal infection in canned compared with other types of foods involved in the incidents. Canned foods accounted for 285 outbreaks out of just over 2800 incidents. Staphylococcal infection in canned meats was 66%, fish 75% and vegetables 80%, whereas it was 23% for all other foods.

Stersky *et al.* [13] stated that in many of the cases only one can was involved. In the majority of the incidents reviewed, information on the condition of the cans or circumstances at the manufacturing plant was not available. Of the 34 incidents where details of the can faults were provided, 19 were related to seam defects which in most cases would not have been picked out by visual examination.

At first sight, the association of *C. botulinum* with leaker infection might not be unexpected. Stersky *et al.* [13] listed three cases involving canned fish packs with *C. botulinum* type E. In their opinion type E botulism from canned foods strongly indicated that leaker infection was the cause of such poisoning and that the frequency of this organism in raw rish accounted for the source of contamination in the canneries. The first incident was in 1934 when one person in Germany died after eating canned sprats. The can was reported as being bulged. In 1963, two women in the United States died after eating a meal including canned tuna. Records showed the batch of cans were properly heat processed [25], but further examinations revealed that 0.5% of cans covering six codes produced at the factory over three months were abnormal. The abnormalities were mainly defective canner's end closures. Many of these cans were found to contain viable micro-organisms and 22 of the cans contained *C. botulinum* type E. In addition, *C. botulinum* type E was recovered from four sites on the can handling equipment which was in contact with cans after heat processing and from raw fish in the plant. However, the NFPA/CMI report [20] states that the cause of the incident was not proven. Schaffner [26] referred briefly to another case involving canned tuna and *C. botulinum* type C and stated a clinical incident had been avoided by the discovery of faulty cans.

The third case of *C. botulinum* type E leaker infection reported by Stersky *et al.* [13] involved $7\frac{3}{4}$ oz Alaskan canned salmon and caused two deaths in the United Kingdom in 1978. The organism gained entry to the can after the heat process through a small hole in the rim of the seam at one end of the can. The cause of the fault was not established although some suggestions were put forward. Little information has been published on the operating practices at the manufacturing plant, but it is known it was the practice to dry protective clothing worn by operatives working in the preparation area by spreading them over the stacks of cans which were air cooled after the heat process.

In 1982 another incident involving $7\frac{3}{4}$ oz Alaskan canned salmon caused one death from *C. botulinum* type E toxin in Belgium. In this case, the can defect, at the time termed a cut-down body flange (CDF)* caused by faulty operation of can reforming equipment at the cannery, resulted in a triangular hole (base 7.8 mm and height 4.7 mm) in the side wall of the can [27, 28]. This incident caused a ban on the sales of $7\frac{3}{4}$ oz canned salmon in the United Kingdom and considerable disruption of trade elsewhere, as stated previously. Examination of quarantined stocks in both the United States and the United Kingdom revealed further examples of the turned down flange fault and a wide range of other can defects. No other can containing *C. botulinum* type E or its toxin was found but other non-toxogenic organisms were recovered from samples of defective cans examined during the investigation. Positive cultures were obtained from about 15% of the defective cans examined (Baird-Parker, 1982, private communication) and from 17.1% of another batch of defective cans examined by Campden Research Association. A report on the latter work including data on the spoilage levels for cans with different types of defects was submitted to the DHSS.

Investigations of some of the incidents occurring in the United Kingdom reported both epidemiological and technical findings in some depth. Couper *et al.* [30], reporting on the typhoid outbreak at Pickering in 1954, were the first to link *S. typhi* with microleakage of cans in untreated cooling water. As a result, a chlorination system was installed at the cannery but the system broke down in December 1962 and was not repaired [31]. The corned beef involved in the typhoid outbreak at Aberdeen in 1964 was packed in 1963 and early 1964 at the same factory. The 1964 Aberdeen incident resulted in the retrospective evaluation of epidemiological and laboratory data, which pointed to other outbreaks from cans infected with *S. typhi* also probably being contaminated through the use of unchlorinated water. In addition, *S. typhi* was found to thrive well in experimentally inoculated cans of corned beef, both on its own and when mixed with *Escherichia coli* and *Enterobacter cloacae*. Under the latter conditions *S. typhi* exhibited strong selective advantages over the other organisms [18].

The investigations of the 1964 Aberdeen typhoid outbreak [17, 31–33] and the 1957 staphylococcal poisoning from processed peas [12] provided a fresh insight into leaker infection and the impetus within the United Kingdom canning industry for a series of further practical studies on the requirements to control leaker spoilage, discussed later. The 1957 staphylococcal food poisoning incidents have resulted in the virtual eradication in the United Kingdom of the practice of manual handling of wet cans after processing and

*Thorpe and Barker (Campden Technical Manual, No. 10) renamed the fault as a turned down flange.

thus removed the most likely cause of food poisoning from leaker infection in this country. This may not be the situation in some countries where the quality of the water used for can cooling may be more variable.

8.6 Blown cans and flat can spoilage

When leaker infection outbreaks occur, the incidence of blown cans is generally low [34]. Bashford and Herbert [22] quoted up to 3% while Segner [8] stated that spoilage levels seldom exceed 1% or 2% even when obvious significant container defects could be shown.

Prior to the 1957 staphylococcal food poisoning outbreaks from processed peas it was generally believed by the UK canning industry that leaking and infected cans, irrespective of the leakage pathway, became blown and therefore could be removed before they reached the point of sale. Gillespy [34] stated that this general assumption was probably true when cooling water was not disinfected. However, with chlorination most leaking cans escaped infection during cooling. He suggested any infection of wet cans at a later stage was more likely to be associated with non-gas forming organisms.

In the 1950s, customer complaints of off-flavours or sourness on opening apparently sound (flat) cans were recognised and taken seriously by vegetable canners in the United Kingdom. These spoiled cans were thought to be due to inadequate cooling and subsequent germination of spores of the thermophilic flat souring organism, *Bacillus stearothermophilus*, that had survived the heat process. This could not be substantiated by microbiological examination of the contents of complaint cans. Frequently non-heat resistant organisms were recovered but were considered to have come from infection of the can contents after opening. Gillespy [34], following the investigation of incidents of staphylococcal infection in processed peas [12] suggested that there was a high probability that many flat sour complaint cans could be attributed to leakage after processing. This opinion was strongly supported by the results of the microbiological examination of 14 600 cans of processed and garden peas discussed previously.

Thorpe and Everton [4] stated that infection by leakage did not always result in blown cans. They also stated the non-blown or flat leaker spoilage level often equalled the blown can level and exceeded it for some products. A study of the reported leaker spoilage food poisoning cases clearly shows flat leaker spoilage can arise from infection with non-gas forming micro-organisms. Bashford et al. [12] reported the examination of stocks of A10 cans of processed peas recovered after the food poisoning outbreaks. Out of a total of 3071 cans, produced on six different occasions between March and September, 71 were found to contain mesophilic organisms; infected cans occurring in every production batch. Four out of the 71 cultures contained *Staph. aureus* phage type 42E, which had been isolated from the incriminated cans and the patients. During July and August the cannery also packed A10 cans of garden peas. There were 2.3% blown cans in the stocks of garden peas. This spoilage rate was exceptionally high and compared with 0.1% blown cans amongst the processed peas. On examination, 14 out of the 91 blown cans of garden peas were found to contain *Staph. aureus* phage type 42E.

The 1964 Aberdeen typhoid incident provided more evidence that infecting micro-organisms do not always result in blown cans and that *S. typhi* and coliforms, which normally produce gas in laboratory media, would grow in cans of corned beef without gas production. Three other typhoid incidents in the United Kingdom in 1963 at Harlow, South Shields and Bedford also showed that *S. typhi* could grow in canned corned beef without gas production. On examination of recalled stocks from the 1963 incidents, it was found that the spoilage level (blown cans) was four times greater than normal. No salmonellae were recovered from these cans [33].

Further evidence of flat leaker spoilage is provided by the incidents of leaker infection involving *C. botulinum* type E in canned tuna in 1963 and canned salmon in 1978 and 1982. In the last incident, the can defect caused a sizeable triangular hole, as discussed previously. While the hole was reported to be 'plugged' with salmon and there was no visible leakage [27], it could be contended that a pressure would not develop in such a can. Examination of stocks revealed more flat but infected cans with various degrees of severity of the turned down flange fault and non-blown infected cans with a wide range of other visual can defects.

Published data relating to commercial spoilage with non-food poisoning organisms are very limited. But within the UK canning industry, many processors have experience of instances where examination of stocks, following incidents of blown leaker spoilage, revealed apparently sound (non-blown) cans infected by non-gas forming organisms. In 97 incidents resulting in blown cans, non-gas forming organisms were recovered in 40% of the cases where apparently sound cans from the same batch were examined (Bean, 1983, private communication).

Thorpe and Everton [4] pointed out that some micro-organisms, for example *Staph. aureus* could grow in products without causing any obvious organoleptic changes to the food. Where details are given of the food poisoning cases reviewed by Stersky *et al.* [31], further evidence is provided to support this view.

8.7 Factors involved in leakage and infection

If leaker infection is to occur, two conditions must be met, namely that the can leaks and that when leakage occurs micro-organisms are present in the surrounding environment. With modern can-making and can-closing equipment, it is possible to obtain economic can manufacture and high closure speeds while providing, in a very high production of containers, protection of the contents from microbial infection. However, it is not possible to guarantee complete protection against infection [4].

Bashford and Herbert [22] stated that while many cans that became spoiled by leakage showed identifiable faults, for example holes in the plate or improperly formed closures, others had no structural weaknesses that differentiated them from good, unspoiled cans. They also pointed out there was ample evidence to show that 2–3% of cans with good commercial seams, i.e. seams made to manufacturers' recommended standards, had temporary leaks induced during cooling and for a short time after. If seams are out of specification, or if they are subjected to abuse during passage of the can handling equipment, then the percentage of leaking cans will be greatly increased [35–38]. The occurrence of a temporary leak does not automatically result in infection of the can contents. Bohrer [39] suggested the volume of cooling water passing into a can could be between 1×10^{-2} ml and 1×10^{-6} ml and Bashford et al. [40] demonstrated that the volume of water entering a can through a temporary leak might be as small as 1×10^{-6} ml. The very small size of leaks is clear from the relationship between the incidence of blown cans and number of micro-organisms in the cooling water [36, 41, 42]. The minute quantity of cooling water admitted by a leaking seam presents no hazard, provided it does not contain micro-organisms.

Summary recommendations for post process can handling were published by the National Canners Association (National Food Processors Association) [43]. The great importance of minimising abuse of cans both before and particularly after heat processing, and the part well-designed and constructed can handling equipment can play in reducing abuse, was stressed by Thorpe and Everton [4] and Thorpe and Barker [29]. Increased spoilage levels when cans were subjected to excessive abuse under factory operating conditions have been reported by Smith [36], Braun and Pletchner [37] and Everton and Herbert [38]. While the effect of various simulated abuse conditions using biotest procedures have been demonstrated by Smith [36], Demsey [44] and Put et al. [41, 42]. Put et al. [41] included data provided as a personal communication from J.A. Perigo (1965) which indicated that light repetitive abuse, which produced no visible deformation of the seam, might nevertheless result in a significant incidence of re-infection. The author believes a reasonable analogy may be drawn between the mode of action of a peristaltic pump and a liquid filled channel through a seam which is subjected to repetitive abuse. This may be one way by which micro-organisms gain entry to the can contents. Other possibilities are that organisms are drawn into the can by the partial vacuum created when the contents are cooled or, in the case of motile organisms, that they are able to 'swim' along the channel. To date little work has been published on the mechanism of microbial ingress into food containers and a better understanding of the phenomenon, which would help to combat leaker infection, is still awaited.

The first chance of infection arises during the later stages of cooling when a vacuum has started to develop in the headspace [45]. The liability to infection persists for as long as the can seams remain wet. It has been equally well established that once seams are externally dry, the liability to infection by leakage virtually disappears. Thus the conveying of dry cans over dry can handling systems does not introduce a spoilage hazard [14, 38, 44, 46–48]. However, if dry cans are rewetted or conveyed over wet runways the spoilage hazard is immediately re-introduced [38].

The ideal of eliminating the circumstances leading to leaker infection, i.e. leaks in cans and the presence of contaminating micro-organisms, cannot be achieved under factory operating conditions. However, may of the essential measures to minimise infection by leakage are under the control of the canner and much may be done to approach the ideal [4]. The introduction of the newest types of cans, side-seam welded and two-piece cans, could be expected to reduce the chances of leaker infection. Nevertheless, it should be clearly understood that the use of these newer types of cans does not remove the need to follow the well-established good manufacturing practices (GMP) to minimise leaker infection. These requirements are listed below and then considered in more detail later.

8.8   Good manufacturing practices (GMP) to minimise leaker infection

The GMPs are not given in order of priority. Effective control of leaker spoilage is achieved when all the GMPs are followed.

- Control critical and major faults in the can structure caused by bad manufacture or inadequate cannery closure or mechanical abuse causing permanent deformation
- Maintain seam measurements within approved tolerances
- Use can cooling water of good bacteriological quality and maintain correct disinfection of that water
- Use procedures to accelerate the drying of cans after cooling
- Control of can abuse; this applies to empty, filled unprocessed and particularly to processed cans
- Correctly clean and disinfect can handling equipment used to convey cans after heat processing
- Use correct manufacturing operations and scheduled processing procedures, particularly during pressure cooling
- Avoid manual handling of wet cans after processing
- Educate operatives and other company staff who may need to handle cans after heat processing in the importance of high standards of personal hygiene, especially handwashing
- Segregate preparation and post process areas and personnel to reduce the chances of microbial contamination

8.8.1   Control of gross faults

The quality of cans is dependent both on knowledge and skills of the can manufacturer, who makes the empty container and the canner, who closes the can after filling. It is not in the interests of either party to have cans with critical or major defects, which in most cases provide a permanent leakage pathway, and therefore a very great probability of spoilage or a potential health hazard should the cans reach the consumer. In addition, cans with defects of this severity may cause jams in equipment particularly some types of continuous sterilisers, for example reel and spiral sterilisers. Faulty cans also frequently lose some of their contents during the heat process and cause pollution of the cooling water system, making effective disinfection treatment more difficult to maintain. The presence of lightweight floating cans in the cooling water may well be an indication of faulty leaking cans and the reason for such an occurrence should be established. In crateless retorts, lightweight cans floating in the 'cushion water' during loading of a vessel may result in damage to incoming cans [29].

Pinholes in the tinplate are nowadays almost always detected and removed automatically at the tinplate mills prior to delivery to the can manufacturer [22]. Within the can-making plants, cans are subjected to further checks for pinholes, plate fractures, split flanges and serious seam faults by passing them through automatic testers on the lines. Such testers work by detection of either a pressure differential or light emission and are capable of rejecting a high proportion of faulty cans.

Three-piece can-making lines in the United Kingdom operate at speeds up to 800 cans per minute. Side-seam welded and two-piece cans are made at speeds up to 500 and 700 cans per minute, respectively. Therefore, quality control cannot be by the inspection of individual containers; the control of gross faults depends on both the examination and reaction to the faults found in cans rejected by the in-line testing equipment and of samples taken regularly from the end of the line.

In the 1978 and 1982 food poisoning incidents involving Alaskan canned salmon, flattened can bodies were reformed and one end put on at the canneries before filling. This practice was adopted for economy in transporting empty cans over long distances and it was not usual to incorporate defect detection equipment in the can reforming lines at the canneries. Today however, many Alaskan and other fish canning factories are now using two-piece cans.

The canner is also faced with the possibility of gross defects but generally does not currently have the facilities for applying in-line detection procedures. For the 1982 and subsequent seasons, a proportion of North American salmon plants had installed automatic equipment to detect leaking cans before the heat process. Such equipment cannot remove all defective cans and should only be considered useful in reducing the chances of faulty cans being released by the canner. The setting of the equipment is very important if maximum efficiency for rejection is to be maintained.

Developments in automatic detection of faults giving rise to permanent leaks are being considered by Campden Research Association and have been proposed by Schaffner [26] in the United States. Present quality control procedures are dependent on sampling and inspection as part of the regular tear-down examination programmes for seam faults or out of specification seam measurements which cannot be detected by external examination. It should be realised that gross faults are more likely to be associated with the restarting of seamers after can jams have occurred and with inadequate maintenance of seaming equipment.

To assist in the recognition of such faults many can manufacturers publish seaming manuals [49–51]. Some of the gross faults are illustrated in a publication by The National Food Processors Association [52] who have also produced an audio-visual aid to complement training programmes and can fault recognition. A more extensive range of visual can defects that can result in leaker infection, together with information on the common causes of the faults, was published by Thorpe and Barker [29].

Permanent deformation of the can seams or bodies due to mechanical abuse after labelling and casing presents a leakage hazard and responsible vendors would not knowingly offer such cans for sale. This form of damage may arise in the processors' warehouses but the majority of the damage is known to occur in the subsequent distribution system. The extent of the damage is not always apparent until the cans are uncased. The industry and Environmental Health Officers in the United Kingdom have frequently expressed concern about the sale of such damaged cans because of the possibility of spoiled product being sold to the public.

8.8.2 *Control of seam measurements within approved tolerances*

The importance of control of closures was appreciated from the earliest days of heat processing. Appert [53] in the first edition of his publication on the Art of Preserving, when he was using glass jars, put much emphasis on the point. One of his four principles was given as: 'to cork these vessels with the greatest care because success depends chiefly on the closing.' Describing his procedure, Appert gave the method of closure in great detail and stated that it was a false economy to use cheap and inferior corks which could result in much greater financial losses if the finished packs spoiled.

The quality of can seams is the responsibility both of the can manufacturer and the canner. As indicated previously, quality control cannot be by inspection of individual containers because of the high speeds of manufacture and closure of cans and the inability to detect deviations from approved dimensional specifications without destructive examination of individual containers. Control is achieved by careful setting up of machines and regular detailed examinations of appropriate numbers of samples. Bashford and Herbert [22] stated that for can manufacture, objectivity was ensured by making as many observations as possible by actual measurements and applying trend analysis for machine performance so that corrections could be undertaken before the quality level fell significantly. The application of this approach was subsequently promoted by the can manufacturers and has been adopted widely by UK canners for the control of their own closing operations. Besides these checks on canner's end seam quality, some canners also undertake quality inspection of incoming cans from the manufacturers.

As previously mentioned, assistance in recognition of common can seam faults and the causes of these defects is provided by manufacturers' seam manuals. Similar publications have been produced by lining compound manufacturers [54]. However, in the author's opinion the most useful document on can faults and their causes was published by Oregon State University [55]. Details of seam measurement and tolerances are nowadays not usually given in can manufacturer's seam manuals because with modern seamer setting practices, the optimum settings vary with the type of seamer and the can used. This information is therefore supplied by manufacturers direct to their customers for the specific closing equipment and the containers in use.

Few publications have given information on the actual methods or routes by which micro-organisms gain entry to cans. Put et al. [41], using information provided in a personal communication by J.A. Perigo, described the alternative leakage routes for micro-organisms to pass through the side seam laps of three-piece cans. Davidson and Pflug [56] listed the location of leaks in the 294 out of 764 blown cans that gave a positive result when vacuum leak tested. In an associated paper [9], they concluded that the vacuum leak test did not provide a good measure of leakage through the double seam. Assuming a hole of diameter 0.0001 in, they calculated the ratio of path length to diameter was 40 times greater through a double seam than through a hole in the tinplate of the can body. Ababouch et al. [11] examined Moroccan canned fish and found that, by using the scoring scheme for seam measurements proposed by Davidson and Pflug [56], they obtained close agreement between the seam quality classification suggesting a potential for leakage and the results of microbiological analyses performed on the cans. A joint project between Campden Research Association and Warwick University has involved the investigation of the mechanism of leakage into containers.

Published data on the effect of different can faults or out of specification seam conditions on spoilage levels are also limited. Where information has been published, it has generally been obtained by biotest experiments and not from commercial operating conditions. Shapton and Hindes [14] referred to earlier work by their colleagues which correlated spoilage levels with various wrinkle gradings. Cans with wrinkle grades of 0–1 (100–75% wrinkle free in current terminology) gave blown can spoilage figures of 75 per 1000 cans, whereas those with wrinkle grades of 2–3 (62.5–50.0% wrinkle free) resulted in 178 blown cans per 1000 packed; the latter figure is statistically significantly higher.

Put *et al.* [41, 42] using biotested cans showed that body hook length, overlap and seam tightness were critical. They concluded from this data that if seam tightness was insufficient, it could be compensated by increasing body hook lengths and having maximum overlap. Very short body hooks could be compensated by sufficient overlap and tight seams. But short body hooks and insufficient overlap could not be compensated by having a tight seaming operation. These conclusions are, however, not in accordance with a major UK can manufacturer's statement of the six requirements for an adequate double seam [57]. Put *et al.* [41, 42] also reported on studies involving the performance of four different lining compounds with seams of varying quality and two types of product, fatty and non-fatty. The interpretation of some of their results, particularly in the case of lining compound No. 1 could not be recommended for commercial operations.

It should be appreciated that the quality control procedures discussed in this section cannot prevent infection from temporary microleaks which are most likely to occur when wet cans are conveyed roughly over contaminated surfaces. However, if the can seam control procedures are inadequate and seam component dimensions are near or beyond the accepted limits, then the incidence of microleakage and infection can be expected to be considerably increased.

8.8.3 *Use of can cooling water of good bacteriological quality and the application of correct disinfection procedures to that water*

Water is the prime transport medium that permits micro-organisms to gain access to can contents via leakage pathways in the container and the seams. In the majority of canning operations water is used to cool products as rapidly as possible after the heat process to minimise overcooking and reduce the total cycle time. Air cooling is only used where suitable water supplies are very limited or expensive and the product is not adversely affected by the much slower cooling procedure, for example in many salmon canneries in Alaska. Investigations carried out by Bryan and Morris [45] on a large number of different can sizes indicated that the ratio of bacterial spoilage from air cooling to that from water cooling was one to ten. A further small-scale trial using cans with artificially produced leaks showed similar results.

With water cooling of cans, the first opportunity for leakage and infection arises in the later stages of the cooling cycle when the transition from an internal pressure to a partial vacuum has occured [58]. Such leakage does not automatically results in spoilage or a health hazard unless the water contains significant numbers of micro-organisms.

There are more references to the good manufacturing practice requirements for cooling water than any other GMP. The existence of the relationship between the bacteriological quality of cooling water and leaker spoilage was realised in the early 1930s and numerous workers published their findings. Bashford [59] stated that as far back as 1931 at least one American worker suspected the bacterial content of cooling water would affect the amount of spoilage in canned foods.

Commercial trials on several canned corn products undertaken in 1931 were described by Scott [60] and illustrate the effect of bacteria free water, attained by chlorination, on the blown spoilage levels. Bacterial counts of the untreated cooling water increased from 7000 to 70 000 organisms/ml as the packing season progressed. This was reduced to 0–1 organisms/ml during the six days of the chlorination trial. The blown spoilage figures, for example, for vacuum packed whole kernel corn cooled in chlorinated and non-chlorinated water were 0.46 and 3.44 cans per 1000. The former spoilage figure appears high when related to the bacteriological condition of the cooling water. Hallman [61], who was associated with the trial, stated that the can seams were of good commercial quality. The cannery was producing about 600 cases per hour, therefore it is possible there was some form of simple runway system which contributed to the contamination of the wet cans.

In 1938, Merrill [62] stated that there was growing realisation of the benefit that could be obtained by using water with low bacterial counts for cooling processed cans, even though by present day standards the control of disinfection was poor. He quoted an instance from 1931 relating to A10 cans of peas in which alternative retort loads had been cooled in clean and heavily contaminated water and from trials made during 1936 on corn products. Spoilage data for comparative trials made during the 1937 packing season were also reported. In the 1936 trials, as well as the one described by Scott [60], the water was not recirculated. Cooling was carried out in tanks or canals to which make-up water, generally of good quality, was added. The addition of chlorine or chlorine releasing compounds was not always continued after the initial treatment of the tank contents and reliance was placed on the make-up water (up to 100 gallons per hour [60]) to dilute any build-up in bacterial numbers. The addition of chlorine was crude and the concentration of residual chlorine was said to be erratic and in one case [60] thought to have been responsible for external can corrosion.

Even at this time, the need to minimise the amount of organic matter introduced into the cooling system, particularly from product adhering to the outsides of cans, was appreciated [60, 62]. The effect of the presence of organic matter in chlorinated water on microbial death rates and on cooling water counts was shown by Mercer and Somers [63] and Bashford [64], respectively, while Thorpe and Everton [4] made practical recommendations on the avoidance of contamination of cooling water systems by organic matter.

Bashford [59] provided comparative spoilage data for a range of canned vegetables, amounting to a total of over 40 000 cans, cooled in ordinary water and the same water after chlorination treatment. Very similar results were obtained in trials reported by Shapton and Hindes [14]. Bashford and Herbert [22] stated that the practice of using chlorinated cooling water or water from a reputable local authority source had become universal in the United Kingdom during the 1940s and any risk of food poisoning from use of sewage polluted cooling water had been eliminated. Gillespy [34] reporting on a post-process sanitation survey undertaken in 1959, stated that 22 out of the 23 factories visited were chlorinating their can cooling water. The remaining one was at that time using municipal mains water once, then running it to waste.

Cooling water counts from 11 canneries in the Netherlands, Belgium and Luxembourg with various water sources and different standards of chlorination were given by Put et al. [41]. They concluded that proper chlorination of can cooling water assisted in minimising the number of bacteria in the water held in the double seams of cans on reaching the end of the runway system. However the bacteriological condition of the runway surfaces is more likely to have a greater influence on the level of seam infection (Wilkinson, 1969, private communication). Graves et al. [65] examined can cooling waters from 30 canneries in America between 1974 and 1975. They found total aerobic counts decreased as the residual chlorine increased. Cooling waters with chlorine residuals of less than 1 ppm were about 1.5 times more likely to have total counts above aerobic plate count (APC) standard of 100 organisms per ml than those with chlorine residual above 1 ppm. The incidence of coliforms was related to the total APC; when this increased, coliforms were detected more frequently. Analysis of the data showed there was a relationship between the bacterial count and the type of cooling system. The order of count (range and median) in increasing order for the different systems was retorts, canals and reel and spiral continuous sterilisers. Odlaug and Pflug [66] reported on a survey of cooling water quality carried out at 17 American canneries in 1975 and 1976. They found the total aerobic count was related to the concentration of hypochlorous acid, the type of cooling system and the water temperature.

Several workers used a biotest procedure to compare spoilage figures for cans cooled in contaminated and uncontaminated water but not subsequently abused. Demsey [44] and Blackwood and Kalber [35] found the spoilage in cans cooled in contaminated water increased by factors of nearly 14 and 8, respectively. More detailed work has been reported by Put et al. [41, 42], who in confirming the findings of earlier workers showed that re-infection was primarily influenced by the number of bacteria in the water trapped in the outer part of the seam, the viscosity of the infecting liquid in the outer part of the double seam and the time the organisms were in contact with the seam. It was considered that if cooling water counts were less than 100 organisms/ml and double seam water counts less than 10 000 organisms/ml, excessive re-infection was unlikely in cans with well-constructed double seams unless they were roughly handled. It was also observed that the cooling water infiltration rate, as a function of the number of bacteria passing continuously through the double seam, was much lower than calculated. From this it was concluded that the leakage path through the double seam was not a straight one, i.e. not in one plane, and tended to filter out micro-organisms passing through the aperture.

Usually, water drawn from municipal supplies in the United Kingdom is satisfactory for can cooling purposes as received at the cannery, i.e. as it is discharged from the water authority main. However, where such water is recirculated, for economy reasons, the bacterial population can build up very rapidly and soon make it unsuitable for can cooling purposes unless disinfection precautions are taken. If economic considerations are disregarded, water of acceptable bacteriological quality can be used once to cool cans and then run to waste. However, the water is normally stored in a holding tank and drawn off as required. This situation can be conducive to bacterial multiplication if the water is static for long periods, for example overnight or a weekend or over a public holiday. The chances of contamination are greatly increased if the tank is not adequately covered to prevent pollution by animal, vegetable and avian sources [4]. Leaker spoilage outbreaks have been caused by cooling cans with water from such a system [22]. Therefore, even water used once for cooling and then run to waste should be disinfected.

Water drawn from rivers, canals or surface water catchments is often used as the main cooling water source or a reserve make-up supply. All such water ought to be regarded as heavily infected with micro-organisms with a high possibility of the presence of pathogenic bacteria and therefore must be adequately treated and chlorinated before use for can cooling purposes [67]. Because of heavier organic loads more chlorination is generally required to attain a satisfactory bacteriological condition for water from these sources. The amount of organic matter, depending on climatic conditions, can also vary quite considerably and rapidly.

A variety of methods to remove or reduce the numbers of bacteria in water supplies has been attempted over the years. These have included irradiation, ultrasonics and chemical disinfection, but it is the last procedure which has received most attention, particularly chlorination treatment. Worldwide experience in the canning trade has established that chlorination, correctly applied, affords a simple, inexpensive, efficient and easily applied method for achieving the disinfection of cannery cooling water supplies [4]. Ito and Seeger [68] discussing the effects of germicides on micro-organisms in can cooling waters, refer to the more recent use of chlorine dioxide and give some of the advantages and disadvantages of this and other disinfectants compared with chlorine. Over the past ten years a few canneries in the United Kingdom have changed to the use of chlorine dioxide but the higher operational costs seem to be a barrier to its wider adoption in the near future.

To ensure effective disinfection of cooling water by chlorine, two requirements must be met. The presence of a free chlorine residual in the water and an adequate residence or contact time for the chlorine to inactivate microorganisms present in the water. Early practical studies showed that free residuals of 1–2 ppm chlorine were very adequate to reduce bacterial populations to low numbers [39, 59, 64, 69]. Bashford [59, 64] quoted a contact time of 20–30 min, from which the current practical minimum contact time of 20 min has been adopted. Thorpe and Everton [4] made recommendations based on factory studies for the requirements to ensure adequate contact times in cooling water systems when discussing the design of cooling water recirculation systems to ensure consistent disinfection.

Chemical tests are used to ensure the chlorine dose rate and the free residual chlorine in the water after can cooling has been effected are at the required level. Tests for the latter should be undertaken every 2 h or more frequently during production periods [4]. The same authors pointed out that the effectiveness of chlorination can only be shown by the low incidence of bacteria in the treated water. Therefore, chemical tests alone are not sufficient and regular bacteriological tests must also be undertaken. Early workers [60, 62] had shown greatly reduced spoilage levels when cans were cooled in water with very low counts. Bashford [64] suggested a maximum total aerobic plate count of 50 organisms per ml, but in 1947 increased the figure to 100 organisms per ml. Strasburger *et al.* [46] quotes the same figure based on the National Canners Association recommendations. Gartland [70] and Herbert [71] stated that there was little or no infection of cans at the cooling stage when the cooling water counts were less than 100 organisms per ml. Murray [72] stated that water used in Northern Ireland canneries should be of the same bacteriological standard as that required for drinking water and have less than 100 organisms per ml using 37°C incubation. This standard of a maximum of 100 organisms per ml, which the work of Put *et al.* [41] supports, has been universally adopted by the canning industry. However, the incubation protocol (time and temperature) is not standaridised. Hersom and Hulland [58] point out that some organisms found in water grow poorly or not at all at 37°C, also the majority of organisms present in cooling water have an optimum growth temperature of 20–30°C. Thorpe and Everton [4] recommended incubation at 25°C for five days for aerobic plate counts (APC). The UK DHSS Code of Practice on the Canning of Low-Acid Foods [67] states the APC should be less than 100 organisms per ml after five days incubation at 20–22°C and includes the additional requirement that no coliform organisms should be detected in any samples of 100 ml of cooling water.

Bacteriological examinations for the presence of spores are not widely undertaken as a routine and there are no published recommendations for maximum acceptable numbers in cooling water. Graves *et al.* [65] in their survey of 30 canneries reported that spores of putrefactive anaerobes were recovered in low numbers (< 1 spore per ml) from nearly a quarter of the samples, whereas over half (53.3%) of the samples contained spores of aerobic bacilli. The difference between the two groups of organisms is probably accounted for by the fact that bacillus spores are more resistant to chlorine than clostridial spores [73]. The same workers in a survey of 17 canneries found the number of spores present was independent of the total aerobic count. Generally the numbers of spores was less than three per ml and was not related to sanitiser concentration. Stalker and Thorpe [74] reviewed published data on the occurrence of clostridial spores in cooling water systems and their resistance to chlorine under laboratory conditions and concluded high levels of spores would lose viability within 30 min in adequately chlorinated water. The NFPA/CMI report [20] also concluded that chlorination of cooling water was effective against clostridial spores.

Thorpe and Everton [4] when discussing the interpretation of bacteriological quality control results for post-process sanitation measures stated that doubt could arise if individual counts were appreciably higher than the standard. In their experience the organism most commonly associated with this situation belonged to the genus *Flavobacterium* and appeared as highly pigmented colonies on cultures, particularly when long incubation times (above three days) were used. Flavobacteria tended to form clumps on wet surfaces, particularly in cooling water systems. This characteristic afforded the organisms a high measure of protection against chlorine in the cooling water. A single high count of these organisms was most probably due to the break-up of a free floating clump collected during sampling. It is now considered that repeated isolation of flavobacteria from cultures of cooling water samples indicates the cooling water system requires to be drained and cleaned out.

Everton *et al.* [75] reported that as a result of an investigation of blown leaker infection outbreak in creamed rice pudding, flavobacteria had been recovered from apparently sound (flat) cans of the product, which showed a marked thinning or decrease in consistency after several months storage at room temperature. The authors reported discussions and work with soup canners indicated the organism was capable of causing a similar type of spoilage resulting in loss of viscosity in cream-style soups. They also reported that previous work carried out in conjunction with this Research Association had shown the organisms were unable to grow and multiply in canned vegetable packs.

8.8.4 *The acceleration of drying of cans after heat processing*

Once the seams of processed cans are externally dry, the vulnerability to leakage and infection virtually disappears and the risk of staphylococcal food poisoning from manual handling of wet cans is removed. With mechanical conveying systems, drying cans reduces the area of can handling surfaces which become wet during production periods and therefore need regular cleaning and disinfection to prevent the build-up of bacterial contamination. Thus procedures which accelerate the rate of drying of can surfaces and seam regions are valuable in controlling leaker infection [4, 29].

Smith [36] stated that spoilage records for batch processing operations indicated an advantage for the practice of drying cans in retort crates before unloading and conveying on runways. When discussing can abuse, he stated that the possibility of spoilage resulting from abuse was greatly magnified while cans were wet. An example of a change of can handling practice was given. Drying cans of cream-style corn before passing them through a corn shaker reduced the spoilage by 50%. Strasburger et al. [46] also recorded a case where allowing retort processed cans to dry overnight in the crates before conveying them over contaminated lines significantly reduced the spoilage.

Data from biotest studies were also quoted by Smith [36]. By drying half a batch of cans overnight before allowing them to travel over a zig-zag runway designed to abuse the cans resulted in 4 blown cans per 1000, whereas the other half of the cans which were run over the abuse hazard while still wet with heavily contaminated cooling water, gave a spoilage level of 20 cans per 1000. Demsey [44] reported biotest results for wet and dry cans run through an entire line after cooling in inoculated water. Wet cans run immediately after cooling gave 21 blown cans per 1000. Those dried overnight after cooling and then run down the line had a spoilage level of 7 cans per 1000. Braun and Pletchner [37] gave an example of reduction of spoilage when a company modified the conveying procedure by inverting the cans so that the canner's end seam was uppermost when travelling along a long cable runway to the labelling equipment and dried more rapidly than the maker's end seam adjacent to the cable. It is presumed the maker's end seams were of a better quality than those of the canner's end. An average blown spoilage level of 1.17 spoiled cans per 1000 was reported by Shapton and Hindes [14] when wet cans were run over wet runways. However, when this practice was discontinued and dry cans were run over dry runways, no blown spoilage was recorded in 60 000 cans. Everton and Herbert [38] quoted a blown spoilage level of 10 cans per 1000 when wet cans from retorts were unscrambled and conveyed to a labeller. With dry cans handled through the same system no blown spoilage occurred.

Braun and Pletchner [37] advocated the use of efficient can dryers for continuous steriliser lines, pointing out the dryer should be located as close as possible to the exit from the cooling section, to avoid bacterial growth on runway surfaces and to minimise the effects of can abuse from automatic conveying equipment. In discussing types of tunnel dryers, they stated that dryers where the can was conveyed upright through a current of hot air were not very efficient. Those which allowed the cans to roll over a steam-heated plate were considered to be more effective. The use of high velocity air jets which blew excess water from cans as they were discharged from the exits of continuous sterilisers was also recommended. Braun [76] again stressed the importance of keeping seams dry if spoilage was to be minimised and restated the case for the use of can dryers on continuously operated lines. Thorpe and Everton [4] described three drying systems in commercial use: high velocity cold air directed from a series of plenum chambers onto cans conveyed in a forward rolling motion; compressed air (when spare compressor capacity was available) directed onto the ends and bodies of cans; and rolling can bodies across a steam-heated chamber operating at 130°C and covered with an absorbent cloth. In the last system, compressed air was used to blow water from the can ends before crossing the heated bed.

Braun and Pletchner [37] reported heavy contamination from felt moisture-absorbing pads on steam-heated hotplate dryers that came into contact with the can double seams. Subsequent trials with pads inoculated with the organism isolated from spoiled packs resulted in a 36-fold increase in the spoilage level compared with the original commercial spoilage figures. The author carried out tests in 1969 to ensure there was no build-up of bacterial contamination on the cloth over the heated bed dryers described previously. A litre of buffered peptone water containing 1×10^7 *Aerobacter aerogenes* per ml was distributed across the cloth surface, approximately 4 ft^2 in area. No organisms were recovered from samples of the cloth cultured after 5 min.

Thorpe and Everton [4] stated that the compressed air systems studied were not as effective as the commercially manufactured, high velocity cold air units and that the compressed air could be wet and therefore condensation traps should be fitted to the supply line. They pointed out that if can dryers were not designed, constructed, installed and maintained properly they could themselves become sources of contamination. Some of the earlier home-made units had been difficult to clean, had not been efficient and in some cases had been the cause of spoilage problems. In the case of high velocity cold systems, drying efficiency could be seriously affected if either exit slots of plenum chambers became partially obstructed and were not regularly cleaned or they were not correctly re-aligned after modification of lines to handle a different size of can. The loss of efficiency of drying encountered when high velocity cold air drying systems are used for cans conveyed upright on cable or slatted belt conveyors was discussed by Thorpe and Barker [29].

No can drying system is completely efficient and passing cans through twists prior to entering the dryer to throw off excess water, particularly from can counter-sinks, was recommended. But at the same time it was pointed out that control of the can feed to the twist was important if build-back and abuse of cans was to be avoided. They recommended that the efficiency of any drying system should be assessed by the method used by Shapton and Hindes [14] and stated that in practice cans (equivalent to the current UT size) from which less than 10 mg of water could be recovered from their external surfaces, including the seams, did not wet surfaces of conveying systems over which they were subsequently conveyed.

It should be appreciated that there are some circumstances when apparently dry cans discharge small volumes of water remaining in their seam regions. This generally occurs when cans change speed or direction rapidly. Conveyor surfaces then become wet and bacterial growth can occur if those sections are not cleaned and disinfected as described in Section 8.8.6. The situation described provides the three main conditions for microleakage and infection, i.e. wet seams, microbial contamination and some degree of abuse, which are frequently associated with changes in direction or speed of cans.

While several workers have recommended leaving retort processed cans to dry in crates before unloading, this practice is not always practical or possible. Crates of cans cooled in retorts, especially vertical ones, often show a temperature gradient which can be as great as 11°C [77] and therefore the rate at which cans dry is variable. Shapton and Hindes [14] pointed out that if cans were brought off at too high a temperature there was a chance of thermophilic spoilage. But if they were brought off too cold and took longer to dry there could be difficulties with labelling and external rusting; besides creating the problem of insufficient stacking space. They referred to trials to reduce the drying time by the use of hot post-cooling dips using three proprietary wetting agents to reduce the surface tension of the water. These allowed water to run off the cans more easily and the residue to spread into thin films and evaporate rapidly. Gillespy and Thorpe [48] reported on factory trials with a wetting agent (CRA1) with specially formulated low foam characteristics. Crates of cans were immersed for 10 s in a tank containing a solution of the wetting agent, before tipping and standing on the floor. Treated cans were dry in 3 h, whereas cans in the bottom layers of control crates still showed traces of water after standing overnight. Shapton and Hindes [14] stated that following their work it had become the practice for certain lines to use the dip procedure and hold crates for a minimum of 1 h. Gillespy and Thorpe [48] stressed that for optimum effectiveness and safety in use, the importance of not overcooling cans, tilting the crates to 60° or as great an angle as possible and holding for 15 s to allow water to run out of can counter-sinks, and ensuring the dip tank solution did not become contaminated by maintaining it at 80°C or by the use of a disinfectant. They reported that the treatment did not increase external can corrosion and, most importantly, that tests undertaken by can manufacturers had not shown that the addition of wetting agents to cooling water increased the water leakage rate through double seams of commercial quality, although increased leakage had been found when seams were loose. Where wetting agents were added to recirculated chlorinated cooling water systems, Thorpe [18] noted that some formulations could interfere with the detection of low levels of free residual chlorine when the diethyl-para-phenyl diamine (DPD) method was used.

Everton and Herbert [38] reported that blown can spoilage levels were reduced from 19 to 1 blown can per 1000 when wet cans from a hydrostatic steriliser discharge were collected and dried before being allowed to pass along the normal can conveying system. Failure to eliminate the spoilage completely by drying the cans suggests the possibility of contamination of cans from double seam contact surfaces within the steriliser before they were discharged [4]. Smith [36] reported spoilage caused by build-up of microbial contamination within the pressure cooling sections of reel and spiral continuous sterilisers. Thorpe and Everton [4] also described the installation and operation of internal can dryers, aided by the use of wetting agents, in one manufacturer's hydrostatic steriliser. This system, if properly maintained, has proved very effective in drying cans before they are discharged and thus avoiding the wetting of subsequent can conveying equipment surfaces during production periods (Shapton, 1980, private communication). The maintenance requirements referred to by Shapton are discussed by Thorpe and Barker [29].

8.8.5 *Control of can abuse*

Other than perhaps imperfect manufacture and closure of cans, rough handling or abuse of cans introduces the greatest possibility of leakage and infection. Can abuse is generally associated with temporary or microleakage, whereas other can faults are likely to result in permanent leakage pathways [29].

Basically there are two types of can abuse. The first is impact abuse which generally occurs when cans roll or slide down runways and either knock into each other or protruding sections of the conveyor. If the impact between either two or more cans or the conveyor and cans is severe then the double seams may be flattened or rim dents may occur just below the seam. The second form of abuse is termed pressure abuse and is normally found where cans on conveyors stop moving forward but the conveyor itself continues to run. Cans ride up on each other so that the seam of one can presses into the body of the next can. The damage occurs just under the seam and may extend for as much as 20 or 30% of the circumference, as although the forward motion of the cans has been arrested they tend to rotate against each other.

Evidence of dents indicates an abuse situation and corrective action should be taken. However, Bashford *et al.* [40] have also shown that light repetitive knocking is just as effective in springing open seams temporarily. The reports of a number of workers discussed later indicate that evidence of dents is not always associated with spoilage related to abuse. Several early papers refer to biotest studies where cans with good seams were shown to leak and become infected when subjected to abuse [36, 76]. Blackwood and Kalber [35] investigated spoilage levels for abused and unabused cans cooled in potable and contaminated cooling water and obtained figures of 0 and 8 spoiled cans per 1000 for non-abused cans with the two conditions of cooling and 4 and 7 spoiled cans per 1000 for abused cans cooled in potable and contaminated water. Strasburger *et al.* [46] found in experimental work that rough handling increased the incidence of spoilage even when no obvious dents were found in the cans. Demsey [44] using commercial lines, showed a fast rate of dumping cans onto an unscrambler doubled the spoilage (20 cans per 1000) compared with slow dumping. The speed of dumping was under the control of the operator. The unscrambler itself gave a spoilage level of 26 cans per 1000 but this unexpectedly high level was stated to be due to difficulties in removing cans by hand from the unit's exit track. This resulted in the cans receiving unusually rough treatment. The effect of different angles of pitch on the gravity feed track to a caser was also studied. Pitches of 5°, 7.5° and 10° gave 18, 23 and 22 spoiled cans per 1000. The 10° pitch was the normal operating condition for the equipment. Demsey [44] describing early work stated that can abuse on a line was cumulative and gave spoilage levels for each unit forming the line. Three were identified, the dumper, belt conveyor and caser, as causing the most spoilage. The same paper reported work undertaken five years later involving the comparison of three automatic casing units. No significant difference was found with regard to spoilage between the three casers. In the conclusions, comment was made on the very apparent improvement in the spoilage figures of the commercial biotest runs between the two periods. The level was 14 cans per 1000 in the earlier tests and only 1 per 1000 in the later ones. It was suggested that the results demonstrated that improvements in can seam construction had been made that significantly reduced spoilage levels encountered in post-cooling can handling equipment. No information was given as to whether the same lining compounds and deposition rates were used for containers tested on the two occasions.

More recent laboratory work by Put *et al.* [41, 42] showed that as the force of the impact was increased, the incidence of re-infection increased correspondingly. She also quoted other laboratory work by J.A. Perigo (personal communication of 1965) in which he had found the light and repetitive impacts were cumulative and could lead to an incidence of re-infection similar to that from a single more violent impact. On this basis he concluded that under factory conditions repeated but light abuse could give rise to significant re-infection without any visible deformation of the seam.

With the increased use of automatic can handling systems, which occurred earlier in the United States than in the United Kingdom, workers reported the results of controlled tests which showed abuse from conveying cans over runways increased spoilage levels compared with hand casing as a control. Smith [36] quoted spoilage levels of 7 and 16 cans per 1000, respectively, for packs of peas and whole kernel corn handled over the same line but less than 1 per 1000 for cream-style corn which was hand cased. The same cans were used for all three products. In a second case he reported abuse resulting in severe denting of about 80% of the cans and a blown spoilage level of 6 cans per 1000. Shapton and Hindes [14] reported hand casing reduced spoilage to 0.2 cans per 1000 compared with 6–7 cans per 1000 when similar batches of cans were run over normal can handling equipment. Everton and Herbert [38] found a reduction from 22.6 to 5.4 cans per 1000 when cans that normally passed through a badly adjusted case packer, by-passed the unit. In another example, spoilage fell from 5.5 to 0.012 cans per 1000 when a badly adjusted lowerator was removed from the line. Spoilage outbreaks were also reported with automatically conveyed cans which had been cooled in water of good bacteriological quality and on examination revealed no can or closure defects [37, 76].

Thorpe and Everton [4] considered the design and layout of can handling lines with respect to the control of abuse. They stated that abuse of empty and filled unprocessed cans as well as processed cans could affect the chances of leaker infection. They gave detailed recommendations to limit abuse on different conveying equipment, but concluded units such as retort crate dumpers and unscramblers should be replaced if at all possible.

The control of can abuse is one of the more difficult requirements to achieve in the overall control of leaker infection. With current mechanical conveying systems there is always likely to be some degree of abuse. The trend to lighter tinplate makes the need for control more important as does the modern practice of operating higher speed lines. At high speeds it may be advisable to use multi-lane conveying systems to keep the degree of abuse of cans within acceptable limits [29]. These authors discussed the control of can abuse as part of a detailed study on the hygienic design of can handling equipment. They also discussed the basic requirements for the effective operation of can cooling and handling systems associated with crateless retorts.

8.8.6 *Correct cleaning and disinfection of can handling systems used to convey cans after heat processing*

External surfaces and seam areas of cans cooled in properly chlorinated water should be virtually free from external contamination when discharged from retorts or continuous sterilisers. It might be expected, therefore, that runways and other can conveying systems, which become wet from cooled cans passing over them, would not become infected because of the chlorinated cooling water transferred onto their surfaces. However, Thorpe and Everton [4] showed that the level of free residual chlorine in water transferred from wet cooled cans was not sufficient to prevent bacterial multiplication on wet runways and such wet surfaces provided a suitable environment for rapid bacterial multiplication. Counts exceeded one million organisms per 4 in^2 in less than 4 h; even when the surface was initially clean and conformed to the standard proposed by Strasburger *et al.* [46] of not more than 500 organisms per 4 in^2. Thorpe and Everton [4] discussed the requirements for the sanitation of post-process can conveying systems and gave detailed recommendations which are discussed later in this section.

Although Smith [36] had shown the relationship of spoilage to rough handling on mechanical conveying systems of cans cooled in contaminated water, it was the work of Braun and Pletchner [37] which established the association between contaminated runway surfaces and leaker infection. They reported a number of serious outbreaks of spoilage in products conveyed over automatic handling equipment. The cans had been cooled in water of unquestionable bacteriological quality and the containers and their seams were free from faults. Investigations of these outbreaks revealed heavy bacterial contamination around the can double seams and serious foci of infection on the conveying systems. They stated it was surprising to find that conveyors taking cans away from the cooling tanks were in most cases severely contaminated even though few organisms could be recovered from the cooling water. The sections of the conveyor which were especially heavily infected were identified and it was stated that although these surfaces were thoroughly cleaned and disinfected they soon became heavily re-infected. Braun and Pletchner presented typical data for the contamination on different sections of conveying systems in four factories.

The monitoring of bacterial contamination in the double seam areas of cans conveyed over handling systems of four different canneries [47] confirmed the findings of Braun and Pletchner [37]. These later data strengthened the understanding that bacterial contamination could develop on wet can handling equipment even when the bacteriological quality of the cooling water was satisfactory. In addition, the contamination could be transferred onto the can seams and contribute to leaker infection. Everton and Herbert [38] pointed out the chances of ingress of micro-organisms into leaking cans was related to the density of the microbial population around the site of leakage. This observation was later supported by the results of laboratory experiments undertaken by Put et al. [41]. Everton and Herbert [38] reported they had swabbed can runways in many canneries and found astronomical counts in factories where little attention was paid to sanitation. But more interesting, and surprising, sometimes high counts had been found on equipment which received regular cleaning and by visual assessment appeared adequately cleaned. They also gave data to show the degree of improvement which could be achieved when enlightened management was prepared to improve or introduce cleaning programmes. Further data on the bacterial contamination on conveying systems were given by Put et al. [41] in their survey of 11 Benelux canneries.

As previously mentioned Thorpe and Everton [4] made recommendations on sanitation programmes for can handling systems. These were based on a three year co-operative study with 25 canning companies. Part of the project involved assessing the effectiveness of existing sanitation programmes and then investigating the effect of the introduction of modified ones. They found that heavy infection developed in a relatively short time on all surfaces that became wet during production periods unless these surfaces were cleaned and disinfected at least once every 24 h. However, adoption of such a cleandown programme alone would not maintain the surfaces within the standard of 500 organisms per 4 in^2. To achieve the standard, it was necessary to mist spray the surfaces regularly with a disinfectant during the production periods. Quaternary ammonium compounds were preferred to chlorine solutions because the latter required more frequent application and caused some corrosion problems at the concentration (300 ppm) required for disinfection of wet runways.

Data were presented to show that the introduction of a sanitation programme to a previously uncleaned conveying system, which was dirty and heavily contaminated, generally required at least a week's application before the bacterial counts were consistently within the standard. The ineffectiveness of cleaning with a solution of chlorine alone was confirmed. Regularly maintained lines could be cleaned effectively with the use of a chlorine-based combined detergent sanitiser. But heavily soiled equipment was better treated by separate applications of a detergent followed by a disinfectant. A rapid and unacceptable increase in microbial counts on surfaces was also demonstrated on both experimental and commercial can conveying systems if product debris from damaged cans was not hosed off immediately.

Badly corroded surfaces provided greater adhesion of soils. Micro-organisms growing in pits or crevices were protected from disinfectants and continually 'seeded' the line making it more difficult to maintain the standard. Some materials, for example studded rubber and canvas belts used on certain types of can elevators could not be effectively cleaned and the replacement of such materials was recommended. The relationship between high microbial counts and unsatisfactory design and layout of conveying systems was demonstrated. Runways conveying unprocessed cans could be heavily contaminated and could infect post-process can handling lines from continuous sterilisers where in-feed runways crossed over discharge runways. In addition, it was shown that some types of hydrostatic sterilisers were capable of contaminating cooled cans before they were discharged from the machine and that this contamination could then be transferred to subsequent sections of the conveying system.

The requirement to reduce can abuse on high speed can handling lines has resulted in the introduction in food canning plants of multi-lane slatted belt conveying systems, which were originally designed for handling beverage cans. The author has investigated a number of these installations as a result of leaker spoilage incidents. Bacteriological examination and study of the design and construction of these slatted belt conveyors has identified a number of unsatisfactory aspects with regard to post-process sanitation. Recommendations on these types of conveyors have been included in an overall reassessment of the hygienic design requirements of can handling systems currently in use in the United Kingdom canning industry [29].

8.8.7 *Use of correct manufacturing operations and processing procedures, particularly pressure cooling*

Any circumstances which result in the development of an excessive internal can pressure may strain or permanently distort the container end panel and the surrounding double seam and thus increase the chances of leaker infection occurring. The pressures which develop in a can are dependent on the amount of air present in the can (vacuum), the headspace, the can end cover design and materials and the heat processing temperature. Braun [76] pointed out that various steps in packing procedures had a vital influence on the performance of the double seam and stated that the packing of fruits and vegetables, containing interstitial gases, without adequate exhausting or blanching would stain the seam during the process and could result in seam leakage. Put et al. [41] stated that overfilling cans so that there was no headspace resulted in seam deformation and leakage even when the correct heat process cycle (presumably including pressure cooling) had been applied.

Faulty retort operation resulting in excessive internal pressure in the can particularly at the end of the process, i.e. during cooling, was recognised by Braun [76] as another factor tending to result in double seam leakage. Put et al. [41] in laboratory trials showed that abrupt pressure changes inside or outside the can, which could arise from incorrect retort operation procedures during the heating or cooling cycles, could deform the double seam and lead to increased leaker infection.

The statements and work of the authors mentioned above may be considered theoretical and not likely to result in spoilage outbreaks or food poisoning incidents under commercial operating conditions. Bashford et al. [12], however, in discussing the circumstances of the 1957 staphylococcal food poisoning outbreaks in the United Kingdom, stated that one of the contributory factors was the unsatisfactory pressure cooling of the A10 cans of processed peas which had caused the seams to be stained and this was shown by the peaking of some of the cans.

8.8.8 Avoidance of manual handling of wet cans after processing

The manual unloading of wet cans from retort crates after processing has been shown to have resulted in a number of staphylococcal food poisoning incidents. The most fully documented incident was described by Bashford et al. [12]. The relationship between leaker infection and food poisoning, discussed previously, showed that staphylococcal infection accounted for a very high proportion of the total food poisoning outbreaks from canned foods [13, 23, 24].

Gillespy and Thorpe [48] pointed out that the primary source of staphylococcal infection is nearly always a human carrier and that more than half the population carry staphylococci in the nose. Therefore, it was inevitable that the hands become affected from time to time. In addition, people suffering from boils or septic cuts or burns were even more likely to carry greater numbers of staphylococci on their hands and should not be allowed to handle processed cans.

Gillespy and Thorpe [48] did not support the use of rubber gloves by operatives who unloaded wet cans. They stated that bacteria could multiply very rapidly on hands inside gloves. Therefore, unless the gloves were quite impervious and changed and sterilised at short intervals, contamination of cans from gloved hands might be greater than from bare ones. Gillespy and Thorpe [48] recommended that manual handling of wet cans should be avoided whenever possible, even at the cost of some inconvenience. They suggested the use of wetting agents to speed up drying as discussed in Section 8.8.4. Thorpe and Everton [4] made a more stringent recommendation that cans should not be manually handled at all while still wet after processing. The same requirement is given in the DHSS Code of Practice No. 10 on the Canning of Law-Acid Foods [67].

In the United Kingdom this requirement is generally complied with for the manual unloading of batch processed cans. But it is not always possible to avoid some handling of wet cans on mechanical conveying equipment, for example uprighting of occasional cans that have fallen over or the freeing of jammed cans. Some companies have attempted to introduce the use of tongs or disinfected gloves for operatives to handle such cans. However, in the author's experience the oepratives' first reaction is to clear the cans by hand rather than to use one of the items mentioned above. This is one reason why it is recommended that all operatives working with processed cans should be made aware of the personal hygiene requirements discussed in Section 8.8.9 and the importance of observing these measures.

8.8.9 Education of operatives in the importance of personal hygiene, especially handwashing

The importance of avoiding external contamination of cans while they are still wet has been stressed (section 8.8.8). It has also been pointed out that external contamination of cans with food poisoning organisms, with the risk of subsequent infection by leakage, is most likely to occur when wet cans are handled by operatives.

Thorpe and Everton [4] stated the objectives or personal hygiene measures were to ensure that:

- Operatives' hands are physically clean and as far as possible free from micro-organisms before they handle processed cans.
- Recontamination of hands while at work is reduced to a minimum.

They considered the same standard of personal hygiene should apply to all personnel who handled or might handle cans after processing. Therefore, operatives manning the exits of continuous sterilisers and labelling equipment should not be overlooked. As part of the recommendations for personal hygiene of operatives handling processed cans, they also discussed the requirements for the location and facilities for handwashing.

The same requirement for personal hygiene should also apply to factory management, who do not always set a good example and handle cans themselves or allow visitors to handle cans. Such a practice undermines the effectiveness of company training programmes.

They also stated that the management was responsible for ensuring that all operatives were aware of and understood legal and factory regulations including those concerned with personal hygiene and habits. The regulations should be clearly stated in a company document given to each employee. They recommended that besides this operatives should be instructed by notices, lectures or talks on the importance of personal hygiene and habits and such education should be an ongoing process.

Thorpe and Everton [4] suggested that the following requirements for personal hygiene related to post-process sanitation should be considered as minimal.

- Operatives who are suffering from heavy colds, bowel disorders, or any sty, boil, carbuncle, abscess or septic skin lesion should not be allowed to handle processed cans
- Operatives who have been in contact with infectious diseases, or who are suffering from any of the conditions listed above must report to their supervisor, the factory nursing staff or the manager before starting work
- Operatives must wash their hands before starting work, after tea and meal breaks or any temporary absence from the line and always after visiting the lavatory
- Operatives must wear suitable protective clothing and head coverings, which should be kept clean

The FAO/WHO Codex Code of Practice for Low-Acid and Acidified Low-Acid Canned Foods [79] includes a section on personal hygiene and health requirements for personnel. The DHSS Code of Practice on the Canning of Low-Acid Foods [67] quotes a similar requirement for hygienic standards of personnel and, in addition, recognises the need for specific requirements for personnel involved in post-processing handling of cans.

8.8.10 *Segregation of preparation and post-process areas and personnel*

It is a basic principle of food hygiene that raw and cooked products should be prepared and stored in separate areas. In addition it is good practice to use different employees in the two areas. Both these steps reduce the chances of cross-contamination.

When considering the design of a canning factory and the layout of equipment within the plant it is logical to ensure that different sections of the line (reception, preparation, filling, processing and casing) are clearly separated from each other. Generally, reception to filling areas can be considered as involving wet operations, whereas labelling and casing are dry operations.

Gillespy and Thorpe [48] stated that separation in a canning factory was normally achieved by reasonable spacing of the equipment, but sometimes was by means of physical barriers. They recommended that preliminary preparation operations, for example with root vegetables, should be undertaken outside the manufacturing area to reduce dust (and microbial) contamination in the plant.

The DHSS Code of Practice No. 10 on the Canning of Low-Acid Foods [67] stated that it was desirable to separate processes and listed areas for nine purposes. The extra areas were for separate storage of various packaging materials, waste products and by-product materials.

Thorpe and Everton [4] pointed out that operatives may be moved to and from preparation and post-process can handling areas to meet specific production requirements. For example, at the start of production, operatives normally employed on labelling and casing might be moved temporarily to inspection lines until cans were discharged from continuous sterilisers. When these employees returned to their normal work areas they could cause cross-contamination. It is recommended that such practices are avoided as far as possible. However, if movement of staff from one area to another is necessary care should be taken to ensure that the operatives involved change their protective clothing and footwear and thoroughly wash their hands before returning to the post-process can handling area.

Sterksy et al. [13] discussing incidents of leaker infection in fish packs associated with C. botulinum type E stated that the frequency of this organism in raw fish provided the source of contamination in canning. In the leaker infection incident involving C. botulinum type E in tuna in 1963, Eadie et al. [80] reported the organism had been recovered from scrapings taken from the cannery's assembly line beyond the cooking facility. An anonymous report [81] stated the organism was recovered from four sites on the post-process can handling system. One explanation for the post-process contamination could be that there was insufficient segregation of preparation and post-process handling areas. In the 1978 incident involving leaker infection by C. botulinum type E in canned salmon, it is believed that the one infected can was contaminated by wet protective clothing used by preparation operatives being dried over warm stacks of processed cans which were air cooling. This practice was made possible by the lack of segregation and the close proximity of the two operations.

In 1982 two further cases of botulism occurred as a result of leaker infection in canned salmon packed in 1981. Cross-contamination from the preparation to post-process area could account for the source of infection. However, there is no suggestion that the organism was transferred in the same manner as occurred in the case of the 1978 incident.

References

1. Gillespy, T.G., Cannery hygiene, Technical Bulletin No. 6, Campden Food Preservation Research Association, 1961.
2. Goldblith, S.A., Quality guarantees, Food Technol. Oct. (1972) 40.

3. Gillespy, T.G., Aspects of the control of leaker infection, Technical Memorandum No. 40, Fruit and Vegetable Preservation Research Association, Chipping Campden, Glos. England, 1961.
4. Thorpe, R.H. and Everton, J.R., Post-process sanitation in canneries, Technical Manual No. 1, Campden Food Preservation Research Association, 1968.
5. Subba Rao, M.A., Spoilage in Indian canned foods, *J. Sci. Ind.* **21D** (1962) 427.
6. Kefford, J.F. and Murrell, W.G., Some problems of spoilage in canned foods, *Food Technol. Australia* **7** (1955) 491–498.
7. Richardson, K.C., Microbial spoilage in Australian canned foods 1955–68, *Food Pres. Quart.* **29** (1969) 52–56.
8. Segner, W.P., Mesophilic aerobic sporeforming bacteria in the spoilage of low-acid canned foods, *Food Technol.* **33** (1979) 55–59.
9. Pflug, J., Davidson, P.M. and Holcomb, R.G., Incidence of canned food spoilage at the retail level, *J. Food Prot.* **44** (1981) 682–685.
10. National Meat Canners Association, Field performance of metal food containers with easy-open ends, *Food Technol.* Feb. (1975) 48–49.
11. Ababouch, L., Chonguer, L. and Busta, F.F., Causes of spoilage of thermally processed fish in Morocco, FAO Fisheries Report No. 329 (Suppl.) (1986) 300–310.
12. Bashford, T.E., Gillespy, T.G. and Tomlinson, A.J.H., *Staphylococcal Food Poisoning Associated with Processed Peas*, Fruit and Vegertable Preservation Research Association, Chipping Campden, Glos., England (1960).
13. Stersky, A., Todd, E. and Pivnick, H., Food poisoning associated with post-process leakage (PPL) in canned foods, *J. Food Prot.* **43** (1980) 465–467.
14. Shapton, D.A. and Hindes, W.R., Some aspects of post processing infection, *Proc. 1st International Congr. Food Sci and Technol.*, ed. J.M. Leitch, Gordon & Breach London, 1962, pp. 205–211.
15. Murrell, W.G., The spoilage of canned foods, *Food Technol. Australia* **30** (10) (1978) 381–384.
16. Howie, J.W., Bacterial survival and spread, *Safety in Canned Foods*, The Royal Society of Health, London, 1966.
17. Howie, J.W., Typhoid in Aberdeen (1964), *J. Appl. Bacteriol.* **31** (1968) 171–178.
18. Anderson, E.S. and Hobbs, B.C., Studies of the strain of *Salmonella typhi* responsible for the Aberdeen typhoid outbreak, *Israel J. Med. Sci.* **9** (1973) 162.
19. Anon, *Grocer* 8 March (1980) 5.
20. Anon, Botulism risk from post-processing contamination of commercially canned foods in metal containers, *J. Food Prot.* **47** (1984) 801–816.
21. Todd, E.C.D., Economic loss from foodborne disease and non-illness related recalls because of mishandling by food processors, *J. Food Prot.* **48**(7) (1985) 621–633.
22. Bashford, T.E. and Herbert, D.A., Post-process leakage and its control, *Royal Soc. Health Conference on the Safety of Canned Foods*, 1965.
23. Gilbert, R.J., Kolvin, J.S. and Roberts, D., Canned foods—the problem of food poisoning and spoilage, *Health Hygiene* **4** (1982) 41–47.
24. Cockburn, W.C., Food poisoning (a) reporting and incidence of food poisoning, *The Royal Society of Health, London* **80**(4) (1960) 249–253.
25. Anon, *Canning Trade* **85**(20) (1963) 10.
26. Schaffner, R.M., Government's role in preventing food-borne botulism, *Food Technol.* Dec. (1982) 87–88.
27. Anon, Canned salmon recall—1982, National Food Processors Association Res. Lab. Ann. Report 3, Washington, DC, 1982.
28. Anon, The tin of salmon had but a tiny hole, *FDS Consumer* **16**(5) (1982) 7–9.
29. Thorpe, R.H. and Barker, P.M., Hygienic design of post-process can handling equipment, Technical Manual No. 8, Campden Food Preservation Research Association, 1985.
30. Couper, W.R.M., Newell, K.N. and Payne, D.J.H., An outbreak of typhoid fever associated with canned ox-tongue, *Lancet* i (1956) 1057.
31. Milne, D., The Aberdeen typhoid outbreak 1964, Scottish Home and Health Dept., Cmnd. 2542, HMSO, 1964.
32. Walker, W. The Aberdeen typhoid outbreak of 1964, *Scot. Med. J.* **10** (1965) 466.
33. Ross, J.M., *The Safety of Canned Foods*, The Royal Society of Health, London, 1966.

34. Gillespy, T.G., Surveys of infection in canned processed and garden peas, Technical Memorandum No. 32, Fruit and Vegetable Preservation Research Association, Chipping Campden, Glos. England, 1960.
35. Blackwood, G.W. and Kalber, W.A., Effects of abuse as shown by contaminated cooling water, *Canner*, Sept (1943) 28–30.
36. Smith, C.L., The relationship of spoilage to rough handling and contaminated cooling water, Continental Can Co. Bulletin No. 9, 1946.
37. Braun, O. and Pletchner, W.L. Recontamination of canned foods after cooling, *Cancer* 114(7) (1952) 12.
38. Everton, J.R. and Herbert, D.A., Post-processing infection cannery hygiene control, *Proc. Int. Food Ind. Congr.*, London, 1962, p. 75.
39. Bohrer, C.W., Post retort can handling problems, NCA paper presented at Tri-State Packers Assoc. Meeting, Dec. 3, Baltimore, MD, 1963.
40. Bashford, T.E., Davies, J.T., Liebmann, H. and Perigo, J.A., Evaluation of can seam construction by chemical and biological means, *Proc. 1st Int. Congr. Food Sci. and Technol.*, Vol. 4, ed. J.M. Leitch, Gordon & Breach, London, 1962, pp. 245–251.
41. Put, H.M.C., Van Doren, H., Warner, W.R. and Kruiswijk, J.T.H., The mechanism of microbiological leaker spoilage of canned foods: a review, *J. Appl. Bacteriol.* 35(1) (1972) 7–27.
42. Put, H.M.C., Witvoet, H.J. and Warner, W.R., Mechanism of microbiological leaker spoilage of canned foods—biophysical aspects, *J. Food Prot.* 43 (1980) 488–497.
43. Anon, Recommendations for post processing can handling, *Canning Trade* March 23 (1964).
44. Demsey, J.N., The effect of post-cooling can handling equipment on spoilage rates, *Proc. Tech. Session at 51st Ann. Conv. of National Canners Association*, 1958; reprinted *Convention Issue Inf. Letter* (No. 1666, 30 Jan. 1958), Washington, DC.
45. Bryan, J.M. and Morris, T.N., Practical canning—the cooling of cans, Research Report 178, Food Investigation Board, Dept. of Science & Industrial Research Report, HMSO, 1932.
46. Strasburger, L.V., Gieseker, L.F. and Jewell, F., Sanitation in relation to post process problems, Technical Bulletin No. 102, Nat. Can. Corp. Lab., USA 1957.
47. Bohrer, C.W. and Yesair, J., Bacteriological studies on post-cooling can handling equipment, National Canners Association Inf. Letter No. 1666, 30 Jan., Washington, DC 1958.
48. Gillespy, T.G. and Thorpe, R.H., Accelerated drying of cans in retort crates, Technical Memorandum No. 52, Fruit & Veg. Can & Quick Freezing R.A., Chipping Campden, Glos., England, 1964.
49. Anon, The formation and evaluation of double seams, Metal Box Ltd. Res. Dept. Tech. Com. No. 15 (revised), Reading, England, 1961.
50. Anon, *American Can Company Seam Manual* No. 4800-M, Top Double Seam Investigation and Evaluation of Meat Cans, 1974.
51. Anon, *Double Seam Manual*, Metal Box, Open Top Group, Reading, England, 1978.
52. Anon, Guidelines for the evaluation and disposition of damaged canned food containers, *National Food Processors Association Bull.* 38L 2nd edn., Washington, DC, 1979.
53. Appert, N., *Le Livre de Tous les Ménages, ou L'Art de Conserver, Pendant Plusieurs Années Les Substances Animales et Végétales*, Chez Barrois l'Aire, Paris, 1810.
54. Anon, *Evaluating a Double Seam*, Dewey & Almy Chem. Div. of W.R. Grace, MA, 1951.
55. Anon, *Top Double Seam Manual*, 4th edn., Dept. of Food Sci. and Tech., Oregon State Univ., 1980.
56. Davidson, P.M. and Pflug, I.J., Leakage potential of swelled cans of low-acid foods collected from supermarkets, *J. Food Prot.* 44 (1981) 692–695.
57. Anon, *Double Seam Manual*, Addendum No. 1, Metal Box plc, Reading, England, 1984.
58. Hersom, A. and Hulland, E.C., *Canned Foods: Thermal Processing and Microbiology*, 7th edn., Churchill Livingstone, Edinburgh, 1981.
59. Bashford, T.E., Infected cooling water and its effect on spoilage of canned foods, *Proc. Soc. Appl. Bacteriol.* (1) (1947) 46.
60. Scott, G.C., Cooling tank contamination, *Canning Age* April (1937) 190–191.
61. Hallman, G.V., Methods for cooling processed cans of meat, *Convention Canner* 86(12) Pt 2 (1938) 104–108.
62. Merrill, C.M., Chlorination of cooling water, *Cancer* 86(12) (1938) 67.
63. Mercer, W.A. and Somers, I.I., Chlorine in food plant sanitation, *Advances in Food Research*, Vol. VII, Academic Press, New York, 1957.

64. Bashford, T.E., Hygiene and canning practice, *Food Manuf.* **20** (1945) 313–318.
65. Graves, R.R., Lesniewski, R.S. and Lake, D.E., Bacteriological quality of cannery cooling water, *J. Food Sci.* **42**(5) (1977) 1280–1285.
66. Odlaug, T.E. and Pflug, I.J., Microbiological and sanitiser analysis of water used for cooling containers of food in commercial canning factories in Minnesota and Wisconsin, *J. Food Sci.* **43** (1978) 954–963.
67. DHSS, Food Hygiene Code Of Practice No. 10, *Canning of Low Acid Foods*, HMSO, 1981.
68. Ito, K.A. and Seeger, M.L., Effects of germicides on microorganisms in can cooling water, *J. Food Prot.* **44** (1980) 484–487.
69. Merrill, C.M., *Canning Age* **22** (1941) 386.
70. Gartland, B.J., Chlorination of cannery cooling water. Application to closed systems, *The Crown* **29**(2) (1941) 17.
71. Herbert, D.A., Canned foods, *Community Health* **1** (1970) 321.
72. Murray, J.G. An approach to bacteriological standards, *J. Appl. Bacteriol.* **32** (1969) 123.
73. Odlaug, T.E. and Pflug, I.J., Sporicidal properties of chlorine compounds. Applicability to cooling water for canned foods, *J. Milk Food Technol.* **39**(7) (1976) 493–498.
74. Stalker, R.M. and Thorpe, R.H., The chlorine resistance of *Clostridia* and their resistance in canning cooling water—A review of published data, Technical Note No. 164, Campden Food Preservation Research Association, 1984.
75. Everton, J.R., Bean, P.G. and Bashford, T.E., Spoilage of canned milk products by *Flavobacteria, J. Food Technol.* **3** (1968) 241–247.
76. Braun, O.G., Performance of double seams, *Cancer* **116** (1953) 22, 24–25.
77. Thorpe, R.H., Atherton, D. and Steele, D.A., Canning retorts and their operation, Technical Manual No. 2, Campden Food Preservation Research Association, 1975.
78. Thorpe, R.H., The determination of chlorine residuals in (a) Cannery cooling water, (b) Factory in-plant chlorination systems, Technical Memorandom No. 79, Fruit and Vegetable Preservation Research Association, Chipping Campden, Glos., England, 1967.
79. FAO/WHO Codex Alimentarius—Recommended Internal Code of Practice for Low-Acid and Acidified Low-Acid Canned Foods, CAC/VOL G. Ed. 1, Rome, 1983.
80. Eadie, G.M., Molner, J.G., Solomon, R.J. and Aach, R.D., Type E Botulism. Report of an outbreak in Michigan, *J. Am. Med. Assoc.* **187**(7) (1964) 496–499.
81. Anon, Another look at post-processing can handling, National Canners Association Res. Inf. Letter No. 83, Oct. Washington, DC, 1963.

Further reading

Anon, Chlorinating cooling canals, *Canning Age* **24**(5) (1943) 248.
Anon, Corned beef and salmon consumption statistics, Ministry of Agriculture, Fisheries and Food, HMSO, 1959–1974.
Anon, Prevention of bacterial recontamination of canned food following heat processing, National Canners Association Res. Inf. Letter No. 50, Washington, DC, 1961.
Anon, Botulism investigation, National Food Processors Association Annual Report 1, Washington, DC, 1978.
Blackwood, G.W., Observations on the effect of handling cans, *Canner* Dec. (1942) 5.
Davidson, P.M., Pflug, I.J. and Smith, G.M., Microbiological analysis of food product in swelled cans of low acid foods collected from supermarkets, *J. Food Prot.* **44** (1981) 686–691.
Denny, C.B., Industry's response to problem solving in botulism prevention, *Food Technol.* Dec. (1982) 116–117.
Matsuda, N., Komaki, M. and Ichikawa, R., Cause of microbial spoilage of canned foods analysed during 1968–1980, *Nippon Shokutin Kogyo Gakkaishi* **32**(6) (1985) 444–449.
Troy, V.S. and Folinazzo, J.F., Handling filled cans carefully can cut your can spoilage, *Package Engineering* **7**(9) (1962) 53–58.

9 The effect of heat preservation on product quality

M.N. HALL and R.J. PITHER

The use of heat treatment for the destruction of microorganisms and the preservation of food products is a well established principle.

The extent of thermal processing which a food receives is dependent upon the composition and physical characteristics of the product and is the result of a combination of time and temperature. The process may be applied either within a sealed container, in the case of conventional canning, or prior to packing under aseptic conditions. In the latter case the food is heat sterilised before being introduced into a sterile container and sealed. Whatever the mode or method of heat sterilisation the safety of a heat preserved product is not dependent upon the use of chemical additives or the control of temperature during storage and distribution. From the point of view of microbial spoilage the shelf-life is considered to be indefinite providing pack integrity is maintained. Physicochemical changes occurring during processing and storage are therefore the factors which determine the product quality in terms of both its sensory properties and its provision of nutrients to the consumer. Reactions take place during both the process itself and on subsequent storage. Generally the changes which occur during storage are slow, particularly when compared with those occurring in an equivalent unprocessed material, and it is on this basis that heat preservation is effective in providing materials outside their normal seasons and in a convenient prepared, often formulated, form ready for consumption or reheating and consumption.

The physical and chemical reactions which occur during processing can be desirable or undesirable, are often more significant and certainly occur much more rapidly than those during storage. As previously noted the degree of heat processing varies according to the product. In turn the changes which occur on processing are influenced by the time and temperature of the process and the composition and properties of the food material [1] and its environment [2]. The following sections discuss the types of changes which occur within the process itself and during storage and concentrates on those which impact on the sensory and nutritional quality of the food product.

9.1 Heat preservation and sensory quality

The heat process itself has a major effect upon the quality of a food product and is responsible for a range of changes which take place (Table 9.1). Starch

Table 9.1 The effect of heat processing on sensory quality.

Texture	
Cell membrane damage	Loss of crispness
Cell separation	Loss of firmness
Protein denaturation	Gelling, firming
Starch gelatinisation	Gelling
Colour	
Natural pigment breakdown	Bleaching
	Loss of colour
Maillard reactions	Browning
Others, e.g. vitamin C	Discoloration
Flavour	
Basic flavour	Stable
Volatile loss (scalping oxidation)	Loss of flavour
Volatile formation (Maillard)	Roasted flavour, bitterness
(oxidation)	Rancidity
(pyracaines)	Roasted flavour

gelatinisation and structural protein denaturations have a direct influence on the texture of a food. Heat induced reactions such as the Maillard reaction affect the colour and flavour as well as the nutritional status of the food [3, 4]. One of the most significant reactions however, is oxidation which can occur during the process and throughout subsequent storage. Flavour [5], colour [6] and occasionally textural changes [7, 8] have all been shown to be related to oxidation, although in the majority of cases the exact mechanisms remain unelucidated. Before any oxidative event can take place, contact with molecular oxygen must have occurred at some point in the history of the food, even as part of the biochemistry of the food components or ingredients as living organisms. In general, that which occurs before the heat process is less important than that during or after processing since it is the manipulative and thermal procedures of food production which have the greatest effect on tissue damage and the resultant mixing of cell contents from different materials.

9.1.1 Texture

Texture appears to be relatively stable on storage of heat preserved food materials, however, certain products do appear vulnerable, e.g. canned yellow plums show considerable softening and breakdown on storage [9]. Particular problems can also occur in fruits, notably apricots and peaches, due to mould infection prior to processing and the formation of heat stable pectolytic enzymes which survive the heat process [10]. The fruit is gradually broken down during storage sometimes to the extent that no recognisable structural properties remain. Far more general are the changes which occur during the heat process itself. The tissue damage which occurs during the heat process of plant material is of two types. These are destruction or damage to the

(a)

(b)

Figure 9.1   (a) Broken surface of raw potato showing ungelatinised starch granules and good cell adhesion. (b) Broken surface of potato chip showing cell separation along the middle lamellae.

semipermeable cell membranes, and disruption of the intercellular structures with resultant cell separation, Figure 9.1, [11]. The effects of these types of tissue damage are a loss in cell turgor and cellular adhesion which give rise to loss of crispness and softening of the heat processed product.

Other major influences on the texture of heated foods arise from the denaturation of proteins. Even on relatively mild heating conformational change affecting the tertiary structure of protein can be observed [12]. Denaturation of the proteins may follow. The hydrogen bonds, maintaining the secondary and higher structure of the protein rupture and a predominantly random coil configuration occurs [13]. This can lead to considerable changes in chemical and physical properties of proteins due to losses in solubility, elasticity and flexibility [12, 14]. This mechanism is also that which causes enzyme inactivation and breakdown of proteinaceous toxins and anti-nutrients. Following denaturation the proteins can form stable aggregates which often have significant effects on the food properties [13, 15]. They cause turbidity leading to either a precipitate or gel, which will greatly alter their water holding capacity and can also lead to increased thermal stability, e.g. β-lactoglobulin when linked to K-casein [16]. Further discussion of the behaviour of proteins in heat preserved food is included later.

Starches are the basic reserve carbohydrate form in plants, being polymers of glucose, and are widely used in processed foods as thickeners. Starch gelatinisation commences at a range of temperatures, which corresponds to the solvation of the macromolecules, and is dependent upon the type of starch present. The difference in behaviour can be partly explained by the relative properties of the two components, amylose and amylopectin. Amylose gives an opaque solution which sets to a firm gel when cooled whereas amylopectin gives a viscous translucent paste which retains its fluidity on cooling. In order for gelation to occur starches must be exposed to both heat and water, but even within a single product gelatinisation is uneven and some unaffected grain will remain at temperatures well above the gelatinisation temperature.

9.1.2 *Colour*

The colour of a food product is determined by the state and stability of any natural or added pigments and the development of any coloration during processing and storage.

Natural pigments are generally unstable compounds which are broken down on heating and storage but whose stability is dependent upon many factors. Some of the major classes of natural pigments are given in Table 9.2.

Anthocyanins are fairly heat stable compounds but take part in a wide range of reactions e.g. with ascorbic acid, sugar breakdown products such as hydroxymethyl furfural and other reactive phenolics which bring about their breakdown [17]. Factors which accelerate degradation include high levels of oxygen in the product and storage temperature. Typical colour loss, observed during storage of red fruits such as strawberries [18], results from breakdown directly influenced by storage temperature.

Conversely, anthocyanins can be undesirable in a product and can be produced on thermal treatment of lencoanthocyanidin [19, 20]. They give rise

Table 9.2 Natural pigments in foods.

Class	Example	Occurrence
Flavonoids	Anthocyanins	Red fruits
Carotenoids	β-Carotene	Carrots
	Lycopene	Tomato
Porphyrin	Chlorophylls	Green plants
	Haemoglobin	Meat
		Products
Betalines	Betanin	Beetroot

to defects such as very dark broad beans and red gooseberries. Other problems can occur with anthocyanin pigments due to the formation of metal complexes, for example, the blueing of red fruits and the pinking of pears when exposed to tin [21, 22]. The flavonoid rutin, present in asparagus, can also form a complex with iron causing dark discoloration in lacquered cans where iron dissolution can occur [6] and in which the colourless tin complex is not formed.

Carotenoids are mostly fat soluble and are responsible for yellow, orange and red coloration. They are unsaturated compounds and are therefore susceptible to oxidation giving rise to off-flavour and bleaching. In addition two types of isomerisation can occur, namely, cis-trans isomerisation and epoxide isomerisation, which can give rise to lightening of the colour. The temperature of storage is considered to have a greater effect on the isomerisation than on the heat process itself.

The two major groups of porphyrin based pigments are chlorophylls and the haem compounds, both of which are very sensitive to heat. On processing, chlorophyll is converted to pheophytin with an associated loss of green colour [6]. Several approaches have been taken to try to reduce the colour loss such as adjusting the pH [23, 24] and the use of HTST treatments. In the latter case although improvements were observed immediately after processing these were lost during storage [25].

Betalins are water soluble pigments which are susceptible to oxidation and loss of red colour. Browning of heat preserved beetroot products is an example where residual oxygen in the product or headspace causes the appearance of a chocolate brown colour.

As well as the breakdown of pigments, as discussed above, oxidation and the Maillard reaction can produce colours during the process and storage. Heat processing itself in the presence of oxygen has a major effect on the end product quality and this is demonstrated by the comparison of products packed in plain tinplate cans with the identical material processed in lacquered cans or glass jars. In the plain tinplate container dissolution of the tin during processing removes a major proportion of oxygen from the pack and little is

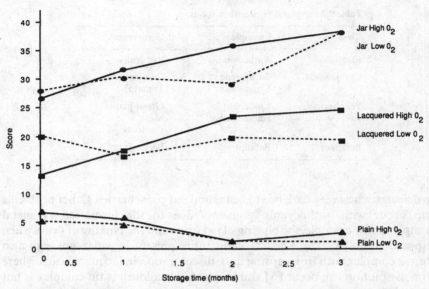

Figure 9.2 Development of brown discoloration in canned beans in tomato sauce.

available to react with the food. Some products such as pale fruits, tomatoes and tomato formulations, mushrooms and milk products are particularly susceptible to such heat induced oxidative changes. Figure 9.2 demonstrates the development of a brownish colour in beans in tomato sauce packed in different container types [26]. Rose and Blundstone showed that the major source of colour change in beans in tomato sauce was the formation of melanoidins from Maillard reaction products described later.

Ascorbic acid is often utilised in products as an antioxidant and can be effective in improving colour in certain products, e.g. mushrooms. It can, however, be degraded to produce reactive compounds which can then further react to form brown pigments.

9.1.3 *Flavour*

Generally, heat preservation does not significantly alter the basic flavours of sweetness, bitterness, acid or salt. Major changes can, however, occur in the volatile flavour components. One of the most important sources of volatiles is lipid oxidation or oxidative rancidity. Lipid oxidation can be brought about during both processing and storage where oxygen is available and is a particular problem in fatty foods and some vegetables notably cereals, legumes and pulses. The proposed chemical mechanisms of oxidation are well documented [5, 27] and are outside the scope of this book. Three stages are involved: i) initiation; ii) propagation in which highly reactive hydroperoxides are formed and iii) termination. The initiation involves uptake of oxygen, in the presence of catalysts, such as metal ions or metalloproteins, but can also be brought about by heat or light. The reaction does, however, have a low activation energy (4–5 kcal mole^{-1}) so that even under low storage temperatures it will continue. The hydroperoxides formed take part in secondary reactions to give rise to a range of volatiles including aldehydes, ketones and alcohols and it is these which produce typical rancid or stale off-flavours.

Volatile flavour compounds are also produced via the Maillard reaction. Since the first scheme for the reaction was put forward by Hodge [28] a great deal of research has been undertaken and the mechanism will not be detailed here. The reaction occurs during heating and extended storage, is influenced by water activity, with an optimum for flavour generation at intermediate values of around 30% water [29] and is accelerated by high pH and buffers such as phosphates and citrates [30]. The first stage of the reaction is fairly well defined and involves the condensation between carbonyl groups of the reducing carbohydrates and the free amino group of the amino acids or protein and rearrangement to produce Amadori compounds. This leads to loss of protein nutritional quality as described later but does not affect the sensory properties significantly [31]. The second stage is very complex and gives rise to numerous products, many volatile and is responsible for many characteristic flavours and off-flavours in food materials. Hodge classified the flavour products of the Maillard reaction into four main groups as given in Table 9.3. The third and final stage of the reaction involves the formation of the brown melanoidin pigments mentioned earlier.

Other volatiles have been identified as having a significant effect on the flavour of foods and perhaps one of the most dramatic is the development of 'catty taint'. This is an extremely unpleasant and potent odour produced by the reaction of unsaturated ketones, notably mesityl oxide, with natural sulphur containing components of the food [32, 33]. Heating is essential in the formation of the taint and incidents have been widespread due to the diverse availability of the unsaturated ketones. Examples include processed meat products using meat from a cold store, painted with a material containing mesityl oxide as a solvent contaminant [34], canned ox tongues which had been hung on hooks coated with a protective oil [35], and pork packed in cans with a side seam lacquer which had been dissolved in an impure solvent [35, 36].

Table 9.3 The major Maillard flavours.

Flavour product	Example	Flavour
Nitrogen heterocyclics	Pyrazines	Nutty roast or baked characteristics
Cyclic enolones	Maltol	Caramel
Polycarbonyls	Pyruvaldehyde	Burnt, pungent aromas
Monocarbonyls	Strecker aldehydes	Aldehydic ketonic aromas

Loss of volatile constituents can also present problems in heat preserved foods. The breakdown of essential oils in citrus products can result from oxidation. Packaging can also have a direct influence on volatile scalping and there is a need for research into the loss of volatiles through modern packaging or absorption by them.

9.2 Assessment of sensory quality

Sensory evaluation can be used to measure either the effects of processing conditions on foods, or their acceptability (like–dislike). The latter tests are carried out by untrained people, the former by trained assessors, usually working under controlled conditions of lighting, temperature, sample size etc. and using well established techniques [37].

The actual test method used is dependent on the purpose of the investigation, e.g. a difference test would be used if the purpose was to find out whether a process had changed a product, a description test to find out how the process had affected it. In all tests those carrying out the evaluations should have been trained in the assessment method.

Appearance, e.g. colour, shape, surface texture and translucency, is assessed visually under standard lighting conditions (e.g. 'Northern light') and against a constant background which will not effect the colour (e.g. gray). Flavour is made up of taste (sweet, salt, bitter, acid) and odour (volatile compounds), it is usually evaluated by month at the temperature at which the food (or drink) is normally consumed. A product can be tasted for total flavour (as in a difference test), as a specific flavour (e.g. a taint) or broken down into the individual flavour attributes in quantitative descriptive analysis (stone and sides), e.g. for coffee, roast, burnt, green acid etc.; the latter two procedures require specific training. Texture can be evaluated visually, e.g. viscosity, by feel e.g. softness of fruit but more generally by mouth when the food is manipulated by tongue, cheeks and teeth. Similar techniques are used to those for flavour analysis of which the most detailed is the Texture Profiling method [38]. A list of some of the terms for appearance, flavour and texture, together with the definition is given by Meilgaard et al. [39].

9.3 Heat preservation and nutrition

The increased demand for convenience foods and the consumption of an ever growing range of product types has meant that heat preserved foods now constitute a significant component of the average diet of individuals in developed countries. The effect that heat preservation has on the nutritional status of foods is, therefore, of considerable importance to food processors, medical practitioners, dieticians and nutritionists, domestic and commercial caterers and every individual as a consumer.

Both physical and chemical reactions occur in heat preserved foods, which influence nutritive value (Table 9.4). Physical factors such as the loss of soluble nutrients, or leaching, can be significant for products in which there is a carrying liquid discarded before consumption. Chemical reactions include heat damage to labile nutrients such as vitamins. When considering the impact of heat preservation on nutritional quality, however, two further consider-ations must be made. These are that the absolute amount of a particular nutrient is often less important that its availability for use by the body and that any comparisons with a 'fresh' equivalent must be made at the point of consumption. Many studies have failed to make appropriate allowance for the degradation which occurs during the storage, preparation and cooking of the fresh material [40]. Only when such allowances have been made can true comparisons with the reheated heat preserved food be drawn.

One of the most fundamental changes which can occur in a heat preserved product is the movement of water and solids within the food material during processing, storage and reheating. In a formulated product or a product in which the entire pack contents are consumed such changes can be largely disregarded, from the nutritional point of view, in that they do not alter the total amount of the nutrients consumed. Products which are packed in a liquor which is discarded before consumption, however, often exhibit dilution, dehydration or loss of total solid materials from the edible portion. In such

Table 9.4 The effect of heat processing on the major nutritional components.

Nutrient	Effect
Dry matter	Loss of total solids into canning liquor Dilution Dehydration
Protein	Enzymic inactivation Loss of certain essential amino acids Loss of digestibility Improved digestibility
Carbohydrate	Starch gelatinisation and increased digestibility No apparent change in content of carbohydrate
Dietary fibre	Generally no loss of physiological value
Lipids	Conversion of cis fatty acids to trans by oxidation Loss of essential fatty acid activity
Water soluble vitamins	Large losses of vitamins C and B_1 due to leaching and heat degradation Increased bioavailability of biotin and niacin due to enzyme inactivation
Fat soluble vitamins	Mainly heat stable Losses due to oxidation of lipids
Minerals	Losses due to leaching Possible increase in sodium and calcium levels by uptake from canning liquor

cases the interpretation of any apparent changes in composition needs to be made with these considerations in mind. From the consumer's viewpoint data expressed on a wet weight basis is likely to be more useful in that this represents the quantities of nutrients available in a serving. The scientist, however, may well be inclined to analyse dry weight data in order to remove dilution or concentration effects to allow a clear perception of true changes in nutrients. The major classes of nutrients which are affected by processing are discussed in the following sections.

9.3.1 *Proteins*

Heat preservation can lead to both desirable and undesirable changes in the nutritive quality of proteins. They are susceptible not only to heat but also to oxidation, alkaline environment and to reaction with other food constituents such as reducing sugars and lipid oxidative products. The total amount of crude protein generally appears relatively unchanged due to heat processing [41, 42] but can suffer from leaching into the liquid component of some products [43].

The crude protein levels, however, appear to be stable during subsequent storage of canned vegetables [41, 42]. The changes which do occur are associated with tertiary structure and functionality as discussed earlier and chemical changes related to digestibility and amino acid availability (Figure 9.3).

Heat sterilisation of meats leads to a reduction in digestibility of the meat proteins and damages amino acids especially the essential sulphur containing

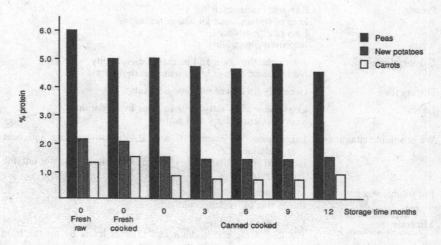

Figure 9.3 Total crude protein levels in fresh and canned peas, new potatoes and carrots.

amino acids and lysine, with 10–15% losses in beef [44]. Canning of potatoes also leads to losses of amino acids though this has been shown to vary depending on the specific gravity of the potato [45]. Lysine is again particularly vulnerable with a reduction in its availability of about 40%. Some of the losses found in canned potatoes, however, may be due to the leaching of the protein into the brine [43] although the major cause of loss of amino acids on heat preservation is the Maillard reaction.

These reactions reduce the protein quality but under normal heat processing are not likely to be significant for most people in developed countries. In certain foods protein availability is improved by heat treatment. Proteins in pasteurised and UHT milk can be better utilised than those in raw milk, as heated milk proteins are precipitated by stomach acids as finely dispersed particles making attack by digestive enzymes easier [46]. Soybeans and many other legumes also undergo improved protein digestibility and bioavailability, especially of the sulphur containing amino acids on heating due to inactivation of trypsin inhibitors and unfolding of the major seed globulins.

9.3.2 Vitamins

The effect of heat preservation on vitamins is generally detrimental although mild heating conditions can have beneficial effects on the bioavailability of certain vitamins, particularly biotin and niacin. This is due to enzyme inactivation and the inactivation of binding agents [47]. The stability of vitamins varies under different conditions with vitamin C and thiamin being most susceptible to degradation through heating (Table 9.5)

Table 9.5 Stability of vitamins under various conditions.

Vitamin	Heat	Air/oxygen	Condition pH 7	Acid	Alkali
A	U	U	S	U	S
B_1	U	U	U	S	U
B_2	U	S	S	S	U
B_6	U	S	S	S	S
B_{12}	S	U	S	S	S
C	U	U	U	S	U
D	U	U	S	—	U
E	U	U	S	S	U
K	S	S	S	U	U
Niacin	S	S	S	S	S
Biotin	U	S	S	S	S
Pantothenic acid	U	S	S	U	U
Folic acid	U	U	U	U	S

U = unsusceptible; S = susceptible
Source: ref [48]

Table 9.6 Vitamin C content (mg/100 g) of fresh and processed peas and carrots.

	Carrots		Garden peas		Canning pea liquor
	Raw	Cooked	Raw	Cooked	
Fresh	9	8	30	16	
Stored 7 days at ambient	3	4	17	16	
Canned	2	3	20	12	11
Canned, stored 6 months	2	2	15	12	10
Canned, stored 12 months	2	3	8	5	5

The fat soluble vitamins are the more stable of the two sets, although these can be degraded by oxidation especially when heated. Losses of water soluble vitamins during processing can be considerably higher.

Vitamin C is the most labile of the vitamins and can be lost during storage of the fresh material, food preparation, washing, and blanching as well as by degradation on heating and leaching into a carrying liquor during the process (Table 9.6). Studies on garden peas and carrots have shown that as much vitamin C can be lost on storage of the fresh produce for 7 days prior to cooking as that lost on canning, although some further losses were seen on extended storage of the canned peas. Much of the vitamin C lost during canning and storage of the canned product is leached into the canning liquor (Table 9.4). Thiamin is the most heat sensitive of the B vitamins especially under alkaline conditions and it is also susceptible to leaching during any washing or blanching stages. Thiamin, however, is less labile than vitamin C and a retention of between 60 and 90% is usual on canning [49].

Folic acid and pyridoxine are also susceptible to degradation by heating and, in the case of folic acid, also by oxidation. Losses of these two vitamins in UHT sterilised milk have been shown to be between 10 and 20% while canning of potatoes can lead to losses of up to 30% [50]. Riboflavin and niacin are both relatively stable on heat preservation although riboflavin is very sensitive to light and will undergo degradation in the presence of both heat and light together [56a].

It is worth restating that the losses occurring in heat preserved foods must be looked at in context with those occurring during storage and home preparation of the food. Heat preserved foods often require less cooking than fresh foods and the differences in vitamin content between fresh and processed food at the point of consumption can often be negligible.

9.3.3 Minerals

Minerals are generally stable to most of the conditions encountered in heat preservation, i.e. heat, air/oxygen, acid or alkali. Losses of minerals, however, can occur during processing, especially of vegetables, due to leaching into the

Table 9.7 Ash content of selected vegetables before and after canning and canned storage.

	% Ash content as consumed		
Sample	Peas	Carrots	Potatoes
Fresh	0.69	0.83	0.76
Canned and stored:			
Zero time	0.99	1.4	1.0
3 months	0.95	1.4	1.0
6 months	0.84	1.3	1.0
9 months	0.99	1.4	0.94
12 months	0.98	1.2	1.10

canning liquor. Conversely, certain minerals, for instance sodium and calcium, can be taken up by the food from the cooking or canning liquids.

Comparisons between fresh and canned vegetables have shown higher ash contents in the canned product (Table 9.7) [41, 42, 50]. This was shown in all cases to be due to the uptake of sodium, and to a lesser extent calcium, from the brine. Storage of the canned vegetables for up to 12 months did not lead to any further major changes in either sodium or calcium content. Between 15 and 50% of potassium can be lost primarily by leaching, on the canning of vegetables. Further losses occur on storage of the heat preserved product but these are not as great as the initial losses. Slight leaching of zinc and negligible changes in iron content occurs during processing with no further loss occurring on storage (Table 9.8). Processing has been seen to increase the bioavailability of iron in spinach and the presence of fructose also leads to increased iron bioavailability [51].

9.3.4 Carbohydrates

Carbohydrates are less susceptible than most other food compounds to chemical changes during heat preservation. The levels of total and available

Table 9.8 Mineral content (mg/100 g) in freshly cooked and canned cooked peas on a wet weight basis.

	Ca	Na	K	Zn	Fe
Fresh	48	65	179	0.82	1.4
Zero time canned	47	320	152	1.0	1.4
Canned stored:					
3 months	40	315	79	0.72	1.3
6 months	31	—	82	0.44	0.9
9 months	28	295	84	0.53	1.5
12 months	—	280	108	0.55	1.2

carbohydrate in vegetables have been found to be very stable on canning and subsequent storage of the canned vegetables [41, 42].

However, there are some effects of heat on the various carbohydrates. The effects of sugars on protein and iron bioavailability and the relationship between starch, texture and palatability have been discussed earlier. Gelatinis-ation of the starch also aids digestibility of foods. A good example of this is the potato which in the raw state is largely indigestible [50, 52]. The exact effect of heat preservation on the various types and constituents of dietary fibre has not been fully investigated. Cellulose, the main constituent of dietary fibre, hemicelluloses and pectins are together responsible for structure and texture in plant foods [53–55] and can be disrupted by heating which leads to a softening of the food and increased palatability as discussed earlier, generally without any loss in the physiological value of the dietary fibre. Storage of canned peas and potatoes showed a loss of fibre after 12 months although the initial canning process had had little effect on crude fibre levels [41, 42]. Overheating can lead to breakdown in the cell walls enabling water soluble nutrients, for instance certain minerals, vitamins and pectins, to be leached out. Although dietary fibre is considered to be largely unaffected by heat processing the exact relationship between time/temperature conditions, dietary fibre breakdown and the extent of nutrient loss due to fibre breakdown requires further study.

9.3.5 *Lipids*

Lipids, especially the unsaturated lipids, are prone to oxidation when heated in the presence of air or oxygen, resulting in losses in nutritional value of the food product. Although the major effect of lipid oxidation is in the flavours of foods, oxidation can lead to a conversion of the natural cis fatty acids to trans fatty acids [51].

The digestion and absorption of trans fatty acids is comparable to that of the cis fatty acids and their nutritional value as an energy source is not affected. However, trans fatty acids do not generally possess essential fatty acid activity, i.e. as precursors of prostuglandins, thromboxanes. This activity is dependent on a cis 9, cis 12 methylene interrupted double bond system but provided that sufficient linoleic acid is consumed (in the UK the average linoleic acid consumption is 4 times the minimum required) the trans fatty acids do not appear to inhibit essential fatty acid metabolism [56, 57].

The oxidation of lipids has also been implicated as previously noted, in the loss of protein quality and can inhibit the activity of the fat soluble vitamins A, D and E as well as vitamins C and foliate. The oxidation of fats in processed foods, however, can be controlled by the exclusion or minimisation of oxygen and the use of antioxidants. The effects of heat preservation on the nutritional value of fats can therefore generally be considered as negligible.

9.4 Conclusions

When considering the effect of heat preservation on the quality of foods two important points should be made:

(i) Many of the changes which occur, both of a sensory or nutritional nature, do so during the thermal process and are not restricted to heat preserved foods. In many instances the process replaces the conventional cooking which a food material receives prior to consumption. Reheating the heat preserved food is a relatively mild treatment which does not significantly affect the quality.

(ii) Heat preserved foods make available to the consumer a wider choice of sensory experience and nutritional requirements without the constraint of seasonality and the burden of preparation.

References

1. Niederauer, T., The influence of technological processes on the nutrient value of foods. *Riechstoffe, Aromen, Kosmetica* **29**(6) (1979) 118–120.
2. Fennema, O., Chemical changes in food during processing. In: *Chemical changes in food during processing*, eds. Richardson and Finley, 1985, pp. 1–14.
3. Mauran, J., Effects of processing on proteins, and Food processing and nutrition: an overview, *Proceedings of the XIII International Congress of Nutrition*, eds. Taylor and Jenkins, 1985, pp. 762–785.
4. Hurrell, R.F. and Carpenter, K.J., Maillard reaction in food. In: *Physical, chemical and biological changes in food caused by thermal processing*, eds. Møyem and Kvale, 1977, pp. 168–184.
5. Hamilton, R.J., The chemistry of rancidity in foods. In: *Rancidity in foods*, eds. Allen and Hamilton, Elsevier Applied Science, 1989, pp. 1–21.
6. Woolfe, M.L., Pigments. In: *Effects of heating on foodstuffs*, ed. R.J. Priestley, Applied Science Publishers, 1979, pp. 77–117.
7. Ory, R.L., *et al.*, Oxidation induced changes in food. In: *Chemical changes in food during processing*, eds. Richardson and Finley, AVI Publishing Co., 1985, pp. 205–217.
8. Matz, S.A., Effects of non-enzymatic chemical changes. In: *Changes in Food Texture*, AVI Publishing Co., 1962b, pp. 262–267.
9. Hall, M.N., Jewell, J. and Henshall, J.D., Shelf-life of canned fruits and vegetables. Technical Memo No. 267, Campden Food and Drink Research Association, 1982.
10. Strand, L.L., *et al.*, Bimodal heat stability curves of fungal pectolytic enzymes and their implication for softening of canned apricots. *J. Food Sci.* **46** (1981) 498–505.
11. Matz, S.A., Blanching, cooking and canning. In: *Food Texture*, AVI Publishing Co., 1962a, pp. 177–191.
12. Finley, J.W., Environmental effects on protein quality. In: *Chemical changes in food during processing*, eds. Richardson and Finley, 1985, pp. 443–476.
13. Ledward, D.A., Proteins. In: *Effects of heating on foodstuffs*, ed. J.R. Priestley, Applied Science, London, 1979, pp. 1–34.
14. Bender, A.E., *Food processing and nutrition*. Academic Press, London, New York, 1978.
15. Tombs, M.P., In: *Proteins as human food*, ed. Lawrie, R.A., AVI Publishing Co., 1970, p. 126.
16. Kinsella, J.E., Relationship between structure and functional properties of food proteins. In: *Food proteins*, eds. P.P. Fox, and J.J. Carden, Applied Science, London, 1982, pp. 51–103.
17. Simpson, K., Chemical changes in natural food pigments. In: *Chemical changes in food during processing*, eds. Richardson and Finley, AVI Publishing Co., 1985, pp. 409–439.

18. Hall, M.N., Shelf-life of canned fruit and vegetables. Technical Memo No. 225, Campden Food and Drink Research Association, 1979.
19. Adams, J.B., and Blundstone, H.A.W., *The biochemistry of fruits and their products*, ed. Hulme, A.C., Vol. 2, p. 513. Academic Press, London and New York, 1971.
20. Adams, J.B. and Ongley, M.M., Changes in the polyphenols of red fruits during heat processing. The degradation of anthocyanins in canned fruit. Technical Bulletin No. 23, Campden Food and Drink Research Association, 1972.
21. Chandler, B.V. and Clegg, K.M., Pink discoloration in canned pears I, role of tin in pigment formation, *J. Sci. Food Agric.*, **21** (1970) 315–319.
22. Timberlake, C.T. and Bridie, P., *Anthocyanins in developments in food colours*, ed. J. Wolford, Applied Science, 1980, pp. 115–149.
23. Malecki, G.J., British Patent No. 772,062, 1957.
24. Malecki, G.J., British Patent No 915,429, 1963.
25. Clydesdale, F.M., Chlorophyllase activity in green vegetables with reference to pigment stability in thermal processing. Ph.D. Thesis, Univ. of Mássachusettes, Amherst, 1966.
26. Rose, D.J. and Blundstone, H.A.W., The reproduction of the effects of plain tinplate in other forms of containers. Technical Memo No. 522, Campden Food and Drink Research Association, 1989.
27. Gunstone, T.D., *Chemical properties in the lipo*, eds. Gunstone, Harwood and Padley, Chapman and Hall, London, New York, 1986, pp. 449–484.
28. Hodge, J.E., Chemistry of browning reactions. *J. Agric. Food Chem.* **1**, (1953) 928–43.
29. Wolfram, M.L. and Rooney, C.S., Chemical interactions of amino compounds and sugars VIII. Influence of water. *J. Am. Chem. Soc.* **75** (1953) 5435–5436.
30. Saunders, J. and Jervis, F., The role of buffer salts in non enzymic browning. *J. Sci. Food Agric.* **17** (1966) 245–249.
31. Hurrell, R.F. and Carpenter, K.J., Mechanisms of heat damage in proteins. 4—The reactive lysine content of heat-damaged material as measured in different ways. *Fr. J. Nutr.*, **32** (1974) 589–604.
32. Aylward, F., Coleman, G. and Haisman, D.R., Catty odours in food: The reaction between mesityl oxide and sulphur compounds in foodstuffs. *Chemistry and Industry*, (1967a) 1563.
33. Aylward, F., Coleman, G., and Haisman, D.R., Catty taints in foodstuffs. Tech. Memo No. 71, Campden Food and Drink Research Association, 1967b.
34. Patterson, R.L.S., Catty odours in food: their production in meat stores from mesityl oxide in paint solvents, *Chemistry and Industry* (1968) 584.
35. B.F.M.I.R.A., Catty taints in foods. *Food Trade Review*, (1969) pp. 47–49.
36. Goldenberg, N. and Mathesan, J.R., Off-flavours in foods: a summary of experience 1948–74. *Chemistry and Industry* **13** (1975) 551–557.
37. BS 5929 *Methods for sensory analysis of food Part I. Introduction and general guide to methodology*, British Standard Institution, London, 1980.
38. Szczeskiak, A.S., Brandt, M.A. and Friedman, J.H., Development of standard rating scales for mechanical parameters of texture and correlation between the objective and sensory methods of texture evaluation, *J. Food Sci.* **28**(4) (1963) 397.
39. Meilgaard, M., Cirille, G.V. and Carr, B.T., *Sensory evaluation techniques*, CRC Press, 1987.
40. Howlett, M.C., A critical review of recent literature on the effects of processing on the composition of vegetables. Technical Note No. 174, Campden Food and Drink Research Association, 1987.
41. Hall, M.N., Edwards, M.C., Murphy, M.C. and Pither, R.J., A comparison of the composition of canned, frozen and fresh garden peas as consumed. Technical Memo No. 553, Campden Food and Drink Research Association, 1989a
42. Hall, M.N., Edwards, M.C., Murphy, M.C. and Pither, R.J., A comparison of the composition of canned, frozen and fresh carrots as consumed. Technical Memo No. 553, Campden Food and Drink Research Association, 1989b.
43. Choudhuin Roy, *et al.* Effect of cooking, frying, baking and canning on the nutritive value of potato. *Food Sci.* (*Mysore*) **12** (1963) 253–5.
44. Czerenski, K. and Jarsabek, K., Changes in the biological value during thermal processing of meat under different conditions, measured by means of the fluorodinitrobenzol determination of available lysine. *Przenyol Spozywezy*, **18** (1964) 714.

45. Jaswal, A.S., Effects of various processing methods on free and bound amino acid content of potatoes. *Am. Pot. J.* **50** (1973) 86–95.
46. Renner, E., Nutritional and biochemical characteristics of UHT milk. *Proc. Int. Conf. UHT Processing.* Raleigh, N.C. 1979, pp. 21–52.
47. Bender, A.C., *Effects of food processing on vitamins,* 1985, pp. 786–790.
48. Harris, R.S., The stability of nutrients. In: *Nutritional evaluation of food processing*, eds. R.S. Harris, and E. Kamas, 1987, pp. 1–5.
49. Benterud, A., Vitamin losses during thermal processing. In: *Physical chemical and biological changes in food caused by thermal processing*, eds. Møyem and Kvale, 1977, p. 199.
50. Woolfe, J., The potato in the human diet, 1987, p. 139.
51. Fennema, O., Food processing and nutrition: an overview, *Proceedings of the XIII International Congress of Nutrition*, eds. Taylor and Jenkins, 1985, pp. 762–766.
52. Birch, G., Chem. Phys. and Biol. changes in CHO's induced in thermal processing. In: *Physical, chemical and biological changes in food caused by thermal processing*, eds. Møyem and Kvale, 1977.
53. Greenwood, C.T. and Mann, D.N., Carbohydrates. In: *Effects of heating on foodstuffs*, ed. Priestly, 1979, pp. 35–76.
54. Cottrell, R., *Food Processing—a nutritional perspective in food processing*, Parthenon Publishing Group, 1989, pp. 189–225.
55. Cottrell, R., *Nutrition in catering*. Parthenon Publishing Group, 1987, pp. 1–31.
56. Harwood, *et al.*, Medical and agricultural aspects of lipids. In: *The Lipid Handbook*, eds. Gunstone, Harwood and Padley, 1986, pp. 527–530.
56a. Priestley, R.J., Vitamins. In: *Effects of heating a foodstuff* ed. R.J. Priestley, 1979, p. 121.
57. Trans Fatty Acids, BNF Task Force report.

Further reading

Dudek, J.A., Elkins, E.R., Chin, H. and Hagen, R. Investigations to determine the nutrient content of selected fruits and vegetables—raw, processed and prepared. *National Food Processors Association*, New York, 1982.
Feeny, R.E. *et al.* Chemical reactions of proteins. In: *Chemical changes in food during processing*, eds. Richardson and Finley, 1985, p 255–283.
Hall, M.N., Edwards, M.C., Murphy M.C. and Pither, R.J., A companson of the composition of canned and fresh new potatoes as consumed, Technical Memo No. 553, Campden Food Research Association, 1989.
Stone, H. and Sidel, J.L., *Sensory evaluation practices*, Academic Press, 1985.

10 Recommendations for the good manufacturing practice of heat preserved foods

J.A.G. REES

10.1 Introduction

The object of preserving foodstuffs is to arrest those natural deteriorative processes, chemical and microbiological, which if allowed to progress would result in spoilage.

In the preservation of foodstuffs by heat, the primary objective is to ensure microbiological stability of the product. This stability is usually achieved by placing the foodstuff in a container, the closure of which will ultimately be hermetic, and after closing submitting the container to a thermal process that will destroy or render inactive all micro-organisms in the product capable of causing food poisoning or spoilage under expected conditions of storage. In aseptic processing, the foodstuff and container may be heated separately and the container subsequently filled and closed under aseptic conditions.

In the United States, regulations for good manufacturing practice in the production of thermally processed foods are in force. Internationally, the Codex Alimentarius *Hygienic Practice for Low Acid Canned Foods*, is scheduled to be published along similar lines. In the United Kingdom specifically, Code of Practice No. 10, *The Canning of Low Acid Foods*, has been issued by the Department of Health and Social Security and is in the process of revision. EEC Directive 77/99/EEC on the *Production and Marketing of Meat Products* contains a chapter specific to hermetically sealed containers. Food processors should be familiar with these recommendations to ensure that the requirements for good manufacturing practice are met.

10.2 Container specifications

(1) Cans and container components used for the preservation of foodstuffs are manufactured from tinplate, aluminium or tin-free steel of specified thickness, temper and where appropriate, tincoating. Three-piece cans have side seams which may be bonded with solder but are usually welded. The solder is of a tin/lead or pure tin composition. The ends of cans are manufactured from tinplate, tin-free steel or aluminium. Can bodies may be straight walled or circumferentially beaded to preserve body strength. Cans may also be of two-piece construction where the can body is without a side seam.

(2) Organic coating materials are specified by a coding system. These materials are selected to be compatible with specific food product categories.

(3) The food processor and/or contractor must satisfy himself that the container specification for each of his products is agreed and recorded.

(4) Can and end specifications as delivered should be checked from batch ticket details, by the canner, to ensure that they comply with the agreed recommendation for the product.

(5) Information provided by the can manufacturer on sequence batch tickets should be recorded as the cans and ends are used. In this way, any problems which may occur subsequently can be related to container production. Ideally pallets should be used in sequence of manufacture/delivery. Incoming acceptance sampling techniques on deliveries of cans and/or loose ends for both appearance as well as structural control should be applied by the canner. These sampling techniques should be agreed with the can manufacturer.

(6) Where plastic containers are used again the container specification and manufacturer product standards should be checked for incoming acceptance. The same criteria outlined in paragraphs (3), (4) (container and lidding specifications) and (5) for cans should be applied.

10.3 Product specification and control

(1) A particular product must be specified by name or code for correct reference purposes.

(2) The formulation of the product must be one which is established in the sense that no change takes place without due reference to the container manufacturer and implementation of agreed testing. Neglect in observing these precauations can result in a change in product/container compatibility which could affect container corrosion, the product and required shelf-life. Of more importance is that the thermal process could be affected as a result of a change in the rate of heat penetration. This change could lead to under-processing with the consequent risk of spoilage, resulting in economic loss and potential food poisoning.

(3) Critical factors which could affect the rate of heat transfer in the pack or otherwise influence the lethality of the process should be known and specified. These include initial microbiological load of raw material and added ingredients, pretreatment, blanching procedure and times, fill weights, head-space, solids/liquid fill ratios, pH, can-closing temperatures and vacuum levels.

Food additives or contaminants in addition to being present in only those amounts permitted by regulation, should be controlled below those maximum levels sufficient to bring about an adverse change in product/container compatibility or accelerated corrosion.

(4) Water treatment at the plant should be controlled and consistent. The presence of nitrate in the water can result in accelerated corrosion in tinplate

containers. Nitrate levels in water should be controlled to as low a level as possible. In the case of acid products (pH 4.5 or lower), particularly when packed into plain tinplate containers, the nitrate content of the product as packed should be less than 5 mg/kg determined as nitrate nitrogen.

(5) Initial trace metal levels in the product are important and should be checked and recorded. These metals (for example lead, iron and copper) may be present in, or be picked up by, the product from boiling pans, pipelines, etc. before it is filled into the can. The presence of these metals can significantly affect container/product compatibility, e.g. the presence of copper could accelerate steel corrosion with some acidic products. Tests used for trace metal determinations should be standard methods, or should be agreed with the container manufacturer.

10.4 Compatibility testing

(1) Before commercial production takes place with any product, adequate testing should be carried out to determine the appropriate container specification and ensure that shelf-life is satisfactory under anticipated marketing conditions. Product details and container specifications should be on record together with test details. Changes in formulation, however minor, invalidate testing packing results and retesting should be carried out on all product modifications. The container manufacturer should be notified of these packing tests and may wish to become involved particularly to comment on the condition of the containers after storage.

(2) Assessment of shelf-life must take into account both the quality of the product and the condition of the container.

(3) Storage conditions are of particular significance and correlation of these with actual conditions in the market place must be considered. Where storage temperatures of or above 35°C are encountered, severe limitation of shelf-life may occur, particularly where acidic products are involved. Controlled storage and regular inspection of such packs are essential, especially in the export market.

10.5 Control of process

(1) Good process control is essential with the thermal preservation of food in containers. Failure to ensure this control can result in container corrosion leading to severe restriction in shelf-life, economic loss resulting from microbiological spoilage and of more important consequence, the possibility of endangering the public health.

(2) Good conditions of storage of empty containers are necessary. Freedom from dust, insects, damp, condensation problems and excessive heat is essential. Containers should be stored well away from any contaminating

materials which may physically enter the containers or generate off-odours or flavours. Long periods of storage should be avoided as quality deterioration may occur and 3 months after the date of manufacture containers should be thoroughly sampled before use.

(3) Containers must be cleaned before filling. Inversion and the use of air jets and/or water jets to remove dust and foreign bodies is recommended. Plastic containers are usually manufactured such that washing with water jets may not be necessary prior to filling. However, it is important to prevent contamination in the plant.

(4) Specification and control of fill weight is important because of the effect on rate of heat penetration, headspace oxygen, corrosion and container performance. Large headspaces could lead to discoloration and/or corrosion on subsequent storage. Significant volumes of air in the headspace could lead to excessive pressure being developed during processing with the consequent risk of permanent container deformation. Alternatively, very low headspaces may again lead to excessive pressure being developed in the container during processing due in this case to expansion of the product. With the use of steel easy open ends a minimum headspace of 8 mm in the closed can must be maintained to prevent strain on the end during processing.

(5) Target in-can vacuum levels must be known and recorded. With few exceptions air removal from the headspace must be achieved for the following reasons.

- It creates an eventual internal vacuum which holds the can ends in a concave shape. Consequently any distortion from the normal shape may be taken as an indication of spoilage either bacterial or chemical.
- It reduces the internal pressure during processing and thus avoids permanently distorted ends. A minimum vacuum level of at least 12.7 cmHg, measured at room temperature, must be achieved when steel easy open ends are used.
- It reduces the level of residual oxygen which would accelerate corrosion during storage.
- With some products, their thermal processes are dependent upon the maintenance of a high vacuum where heat transfer is obtained by means of water vapour given off by the boiling brine condensing on the surface of the exposed solids. Low vacua result in higher initial temperatures being required to start boiling of the brine. Air removal may be achieved by hot filling, steam flow closing or vacuum seaming.

(6a) Can closing (double seaming) conditions are critical, both first and second operations, and must be controlled and regularly monitored. The closing machines must be properly set and maintained to their individual/type specifications. During production it is essential that the seams produced by each individual seamer station are checked for conformance to the specified standards. The process control system must include the recording of specified measurable parameters as well as continuous visual checking to ensure absence of visual defects. Relevant documents which should be studied are:

- Double seam manual
- Double seam set-up and operating specification
- Seamer operations manual
- Spares list

The critical parameters of the double seam which must be controlled are:

- Seam tightness rating
- Actual overlap
- Body hook butting
- Internal droop

with freedom from visual defect and counter-sink length at least equal to the seam length during seam formation (see Figure 10.1).

(6b) Plastic containers may be closed by double seaming or by foil-based or non-foil-based laminates. Again closing conditions are critical with respect to control of fill weight, headspace condition and air removal and very important with heat-seal closing, the degree of flange contamination. Conditions of seal with respect to:

- Temperature of seal
- Pressure of seal
- Dwell time

must be recorded and documented for each product combination. The container manufacturer should be consulted with respect to recommendations for heat-sealing and procedures for assessing seal integrity. A directory of good seals and seal defects should also be kept. National Food Processors Association, Bulletin 41-L, Flexible Package Integrity Bulletin, should be used to assist in the development of the package integrity directory.

(7) Heat processing conditions are critical and must be specified on record and under control. Fill weight, or drained weight where more appropriate, headspace, pH, initial temperature of the coldest spot of the coldest container at the start of the process, container size, retort come-up time, process type, time and temperature must be recorded for each product and container. Full details of retorting must also be recorded, i.e. batch or continuous, static or rotary (including RPM for each product and container), aseptic, etc. For plastic containers it is important to specify retort loading procedures. It is usual to specify self-supporting racks. These racks prevent container to container contact which minimises the potential for abuse and container failure during the processing cycle when containers have softened and are therefore at their most vulnerable. Self-supporting racks are also to be recommended to prevent the restriction of flow of the retort medium. This is particularly relevant to containers processed under water where a 10 mm gap

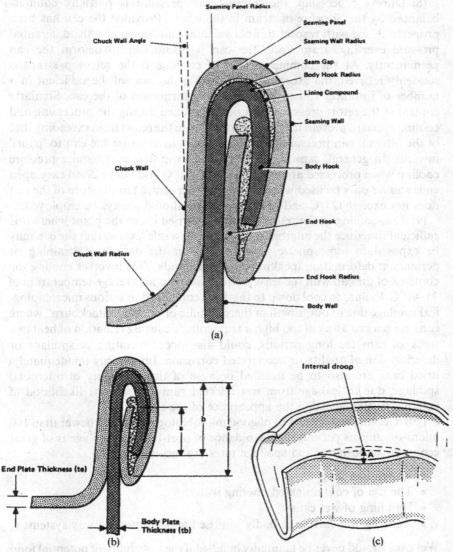

Figure 10.1   Double seaming. (a) General terminology; (b) measurements; (c) internal droop.
(From MB Double Seam Manual.)

between containers is recommended. The process and cooling cycles must be sufficient to achieve commercial sterility. The requirements for the establishment of heat processes and the design and instrumentation of processing vessels are laid out in the USFDA Good Manufacturing Practice Regulations, Title 21; the DHHS Code of Practice and the Codex Alimentarius.

(8) During processing, the internal can pressure is partially counter-balanced by the pressure of steam in the retort. Provided the can has been properly closed with respect to final vacuum and headspace, the differential pressure exerting strain upon the can is insufficient to deform the can permanently. At the commencement of cooling, if the retort pressure is suddenly relieved, the excess pressure within the can will be sufficient in a number of instances to cause permanent deformation of the can. Similarly control of the retort pressure must be maintained during the processing and cooling cycles to prevent the external pressure in the retort from exceeding that of the internal can pressure to such an extent as to cause the can to 'panel' inwards. In general, cans larger than 153 mm in diameter require pressure cooling when processed at temperatures of 110°C and above. Steel easy open ends should only be used where the maximum centre temperature of the can does not exceed 121°C and pressure cooling should always be employed.

(9) Can-cooling in the retort should be carried on to the point where it is sufficient to reduce the internal can pressure to a safe level so that the can may be exposed to atmospheric pressure without the danger of straining or permanent deformation (peaking) of the can ends. This involves cooling the contents of the can with the least possible delay to an average temperature of 35–40°C. Failure to cool down to this level could lead to serious microbiolog-ical spoilage due to outgrowth of thermophilic organisms. 'Stackburn', where cans are packed away at too high a temperature causing retention of heat in a stack of cans for long periods, could also occur resulting in spoilage or deterioration of quality or accelerated corrosion. Imperfectly (inadequately) dried cans are also to be avoided because of the possibility of microbial spoilage, due to leakage from wet infected runways and the likelihood of external rusting affecting the appearance of the cans.

(10) Cooling water must be of good microbiological quality (fewer than 100 micro-organisms per ml). The avoidance of post-process spoilage is of great economic importance and spoilage rates are increased by:

- Rough handling
- The use of contaminated cooling water
- Handling of wet cans
- Running wet cans over badly sanitised and infected runway systems

Wet cans should never be manually handled if the possibility of potential food poisoning through re-infection is to be avoided. This also applies to plastic containers. Where relevant, it is important to maintain effective chlorination of retort cooling water to ensure adequate 'contact time' (minimum 20 min) and the presence of free chlorine. Total chlorine levels should be in the order of 4–5 ppm to avoid can corrosion. The maintenance of an hygienic 'clean-down' and periodic sanitising schedule for all ancillary post-retort equipment is essential. Until the cans are dry, strict control of hygiene must be maintained. Where a chemical disinfectant is used it should be present in the active form after the water has been used as an index of adequate disinfectant prior to use.

Reference to Campden Food and Drink Research Association Technical Manual No. 1 is recommended.

(11) Cans should be dried before labelling and outer packaging and where labels are used, these should comply with acceptable specifications for chloride and sulphate levels, generally agreed as 0.05% as NaCl and 0.2% as Na_2SO_4, respectively.

(12) Containers must be coded so that the filling date can be readily established. With the records of sequence batch information available, as recommended in Section 10.2, paragraph (5), it will be possible to relate filled containers back to empty can/end or container/laminate manufacturing dates if problems occur. However, coding must not be such that it interferes with the integrity of the container closure or organic coating materials. An EEC directive on the requirements for 'lot marking' is in preparation.

10.6 Handling

(1) Containers have limitations in strength; care must be taken in handling pallets of cans to avoid damage.

(2) Conveying and filling of cans requires good line design. Damaged flanges can result in defective seams and abuse can result in corrosion and/or spoilage. Abrasion of the seams may occur in badly designed lines. Score fracture and/or tab removal may occur with easy open ends. The de-nesting of plastic containers, prior to filling, should be such as to prevent permanent distortion.

(3) Filled cans must not be stacked in such a way that the instantaneous or sustained axial (or top) load on individual cans exceeds the safety limits and causes can collapse. Handling in transit must take into account the vulnerability of the can to spoilage if damaged around the seams or the scored easy open end. Regular inspection of filled stacks for spoilage is essential. If spoiled cans are detected in a stack or pallet, the relevant code or codes should be isolated and quarantined pending microbiological examination. Cans with easy open ends fitted should always be stored with the easy open ends uppermost. Adequate protection must be afforded to the easy open end to avoid any pressure or abuse on the panel of the end and general gross abuse. Recommendations for the palletisation of plastics containers should be obtained from the container manufacturer.

(4) Adverse storage conditions, particularly in areas of fluctuating temperatures and high humidities, will promote external corrosion. This will apply to both two-piece and three-piece cans even when the three-piece cans are externally side striped. Codex Alimentarius *Codes of Practice* are proposed to cover:

- Guidelines for the salvage of canned foods exposed to adverse conditions
- Guideline procedures to establish the microbiological causes of spoilage in low-acid and acidified foods

10.7 Packaging and storage for export

(1) The food processor should ensure that the container specification and heat process used in the commercial sterilisation of the product is satisfactory for the market where it will be retailed and that this has been checked by adequate storage testing. The external specification of the can must be tested as well as the consideration of internal product/container compatibility.

(2) Entrapped moisture, such as condensation, should be reduced to a minimum and containers packed as dry as possible. Any contamination with salt water or product from other leaking/spoiled containers must be avoided at all costs. Regular inspection of filled stocks for spoilage/leakage during storage and transit, as indicated in Section 10.5, paragraph (3), is of vital importance in the export of cans to hot climates. Here failure is a greater risk because of the prevailing high temperatures.

(3) A representative sample of containers should be examined prior to despatch for assessment of leakage/spoilage potential, entrapped moisture in-plant damage, etc. Whenever possible, this examination should be conducted at least 72 h after filling.

(4) Overall shrinkwraps should be used if possible on collated containers. Where sleeve wraps are used, they should incorporate board base trays and board 'end' liners. The collated units should be palletised with a chipboard pad placed between the bottom layer of the units and the pallet. The pallet should then be shrouded with an overall shrinkwrapping.

(5) Prior to despatch, storage temperatures should be maintained as constant as possible. Cyclic storage conditions should be avoided. Under normal winter humidity conditions in the United Kingdom, condensation will tend to occur when the temperature of the packs is 7°C lower than the surrounding atmosphere.

(6) Any board used in combination with any external packaging should be compatible with the packs. Low sulphate and chloride levels, as defined in Section 10.5, paragraph 11 are essential. Odour-free materials should also be used.

(7) Unseasoned wood should not be used for pallets.

(8) Palleted cans should, whenever possible, be transported in closed container wagons, particularly when sea passage is involved.

(9) Containers should be inspected on arrival at their destination for leakage/spoilage or damage. Any suspect batches must be isolated immediately pending further investigation.

(10) Prior to and during distribution to retail outlets, storage conditions of the containers should be non-conducive to the promotion of external corrosion. Storage temperatures should be maintained as constant as possible and areas of high humidity avoided in warehousing. Where high humidity conditions exist, maintaining the temperature of the product close to that of the surrounding air will help to minimise condensation.

Index